RENEWABLE FUEL STANDARD ISSUES

RENEWABLE ENERGY: RESEARCH, DEVELOPMENT AND POLICIES

Ethanol and Biofuels: Production, Standards and Potential
Wesley P. Leland (Editor)
2009 ISBN 978-1-60692-224-8

Renewable Energy Grid Integration: The Business of Photovoltaics
Marco H. Balderas (Editor)
2009 ISBN 978-1-60741-324-0

Renewable Fuel Standard Issues
Daniel T. Crowe (Editor)
2009. ISBN 78-1-60692-289-7

RENEWABLE FUEL STANDARD ISSUES

DANIEL T. CROWE
EDITOR

Nova Science Publishers, Inc.
New York

For permission to use material from this book please contact us:
Telephone 631-231-7269; Fax 631-231-8175
Web Site: http://www.novapublishers.com

LIBRARY OF CONGRESS CATALOGING-IN-PUBLICATION DATA
Available upon request
ISBN: 978-1-60692-289-7

Published by Nova Science Publishers, Inc. ✢ New York

CONTENTS

Preface		vii
Chapter 1	Statement of Nathanael Greene Director of Renewable Energy Policy Natural Resources Defense Council Before the Committee on Environment and Public Works Subcommittee on Clean Air and Nuclear Safety United States Senate July 10, 2008	1
Chapter 2	Renewable Energy Policy: Tax Credit, Budget, and Regulatory Issues *Fred Sissine*	29
Chapter 3	Energy Independence and Security Act of 2007: A Summary of Major Provisions *Fred Sissine*	69
Chapter 4	Selected Issues Related to an Expansion of the Renewable Fuel Standard (RFS) *Brent D. Yacobucci and Randy Schnepf*	97
Chapter 5	Robert J. Meyers (Office of Air and Radiation, U.S. Environmental Protection Agency) Before the Committee on Environment and Public Works Subcommittee on Clean Air and Nuclear Safety United States Senate July 10, 2008	125
Chapter 6	Statement of Steven G. Chalk (Renewable Energy, U.S. Department of Energy) Before the Subcommittee on Clean Air and Nuclear Safety Committee	

on Environment and Public Works United States Senate July 10, 2008 **131**

Chapter 7 Statement of John Pierce (Technology DuPont Applied BioSciences E.I. DuPont de Nemours and Company, Inc) Regarding the Renewable Fuels Standard before the Subcommittee on Clean Air and Nuclear Safety Committee on Environment and Public Works U.S. Senate July 10, 2008 **137**

Chapter 8 Testimony of Scott Faber (Federal Affairs On Behalf of the Grocery Manufacturers Association) Before the Senate Committee on Environment and Public Works On "Environmental Protection Agency Oversight: Implementing the Renewable Fuel Standard" July 10, 2008 **147**

Index **157**

PREFACE

High petroleum and gasoline prices, concerns over global climate change, and the desire to promote domestic rural economies have greatly increased interest in biofuels as an alternative to petroleum in the U.S. transportation sector. While recent proposals have set the goal of significantly expanding biofuel supply in the coming decades, questions remain about the ability of the U.S. biofuel industry to meet rapidly increasing demands. This book outlines some of the current supply issues facing biofuels industries, including implications for agricultural feedstocks, infrastructure concerns, energy supply for biofuel production, and fuel price uncertainties. Included in this book is a summary of the Energy Independence and Security Act of 2007 as well as an analysis of this document and testimonial responses to the issues it raises.

Chapter 1 – Biomass feedstocks produced with environmental safeguards, processed efficiently and used in efficient vehicles can reduce our dependence on oil for transportation, reduce emissions of heat trapping carbon dioxide, contribute significantly to a vibrant farm economy, and avoid impacting food prices.

Chapter 2 – High gasoline and natural gas prices have rekindled interest in the role that renewable energy may play in producing electricity, displacing fossil fuel use, and curbing demand for power transmission equipment. Also, worldwide emphasis on environmental problems of air and water pollution and global climate change, the related development of clean-energy technologies in western Europe and Japan, and technology competitiveness may remain important influences on renewable energy policymaking.

Chapter 3 – The Energy Independence and Security Act (P.L. 110-140, H.R. 6) is an omnibus energy policy law that consists mainly of provisions designed to increase energy efficiency and the availability of renewable energy. This report describes the key provisions of the enacted law, summarizes the legislative action on H.R. 6, and provides a summary of the provisions under each of the titles in the law.

Chapter 4 – High petroleum and gasoline prices, concerns over global climate change, and the desire to promote domestic rural economies have greatly increased interest in biofuels as an alternative to petroleum in the U.S. transportation sector. Biofuels, most notably corn ethanol, have grown significantly in the past few years as a component of U.S. motor fuel supply. Ethanol, the most commonly used biofuel, is blended in nearly half of all U.S. gasoline (at the 10% level or lower in most cases). However, current biofuel supply only represents about 4% of total gasoline demand.

Chapter 5 – Renewable fuels are a key element of a national strategy for addressing our energy security and the challenge of global climate change. The national renewable fuel standards, in combination with the vehicle fuel economy standards in EISA, will reduce emissions of greenhouse gases in the transportation sector and improve our energy security. The changes brought about by EISA are expected to prevent the release of billions of metric tons of greenhouse gases emissions into the atmosphere over the next several decades.

Chapter 6 – The implementation of the RFS is one way that Congress and the Administration have recognized and responded to the urgency of expanding use of non-petroleum fuels to improve energy security, reduce costs, and reduce greenhouse gas emissions.

Chapter 7 – DuPont brings a broad perspective to bear on biofuels issues, as we span the biofuels value chain. Our seed business Pioneer Hi-Bred is the world's largest seller of seed corn and the second largest seller of soybean seed to farmers. Our seeds have enabled steadily rising yields, or production per acre, for over 80 years. We sell over 170 corn varieties specifically for ethanol production, varieties that produce high fermentable starch, yielding more gallons of ethanol per acre.

Chapter 8 – Food and beverage companies strongly support the development of sustainable bio-fuels that contribute to America's energy security and that do not pit the nation's energy needs against the needs of the hungry or the environment. We are working with a broad coalition of industry, farm, labor, anti-hunger, consumer, minority, and environmental organizations to urge Congress to revisit our food-to-fuel policies in light of runaway food inflation and new questions and concerns regarding the environmental costs of food-to-fuel production.

In: Renewable Fuel Standard Issues
Editor: Daniel T. Crowe
ISBN 978-1-60692-289-7

Chapter 1

STATEMENT OF NATHANAEL GREENE DIRECTOR OF RENEWABLE ENERGY POLICY NATURAL RESOURCES DEFENSE COUNCIL BEFORE THE COMMITTEE ON ENVIRONMENT AND PUBLIC WORKS SUBCOMMITTEE ON CLEAN AIR AND NUCLEAR SAFETY UNITED STATES SENATE JULY 10, 2008

ABSTRACT

- Biomass feedstocks produced with environmental safeguards, processed efficiently and used in efficient vehicles can reduce our dependence on oil for transportation, reduce emissions of heat - trapping carbon dioxide, contribute significantly to a vibrant farm economy, and avoid impacting food prices.
- Pursued without adequate guidelines, large scale biofuels production carries grave risk to our lands, forests, water, wildlife, public health and climate.
- The current ethanol excise tax lacks any food security or environmental safeguards. Congress should replace the current tax credit, the ethanol import tariff, and the biodiesel blending tax credit with technology - neutral, performance - based incentives that reward the production of biofuels that avoid food disruptions, increase water efficiency, reduce water pollution, improve soil management, and enhance wildlife management.

- The Renewable Fuel Standard contained in Energy Independence and Security Act of 2007 contributed important advances to our energy and climate policy that can help mitigate global warming, reduce the environmental impacts of biofuels, and start to take biofuels out of the food price equation. The latest research confirms Congress' foresight in establishing provisions to:
  - Require conventional biofuels from all new facilities to achieve at least a 20 percent reduction in lifecycle greenhouse gas emissions compared to conventional gasoline and advanced biofuels to achieve at least a 50 percent reduction.
  - Define lifecycle greenhouse gas emissions to include the full cultivation, production, and combustion cycle of fuels and the emissions from direct and indirect land use change caused by this cycle. This critical step helps ensure biofuels produce real climate benefits and provides a direct disincentive to displacing food.
  - Encourage production of plentiful biofuels feedstocks—including woody - biomass—while ensuring the RFS mandate does not result in the loss of old - growth forest, native grasslands, "critically imperiled", "imperiled", "vulnerable" ecosystems pursuant to a State Natural Heritage Program, the degradation of our federal forests [1], or conversion of natural forests on non - federal lands.
  - Require the vast majority of new biofuels required under the law to be advanced biofuels derived from renewable cellulosic biomass with a 60 percent lifecycle greenhouse gas emissions reduction.
  - Establish a no - backsliding requirement to protect air quality by directing EPA to adopt regulations that "mitigate, to the greatest extent achievable ... any adverse impacts on air quality."[2]
- Congress should reject current efforts—like H.R. 5236 and S. 2558—to revoke, repeal, or dilute the renewable biomass protections, as well as any efforts to repeal the GHG standards. These provisions are not only critical to keeping biofuels production from producing negative impacts today, but they also represent the foundation for any successful biofuels policy going forward. New crops and conversion technologies are developing rapidly that will make it easier to produce lots of biofuels with a smaller environmental footprint and without impacts on food prices, but the technologies are not a guarantee of good environmental performance. We need to maintain the environmental safeguards and performance standards in the RFS and build on them, guiding the market so that innovation and competition will drive biofuels to provide the greatest benefits.
- The RFS also includes important requirements for studies of various aspects of current and future biofuels. Seemingly every day now, we learn of new technologies that promise to improve the performance of biofuels and of new negative impacts that biofuels can have if pursued

carelessly. These studies are critical to ensure that we identify unintended consequences of our policies as soon as possible and get the greatest good from our policies.

- Congress should make sure EPA is fully funded to aggressively and effectively implement these critical safeguards and should monitor their progress closely to ensure that science rather than politics drives the resulting regulations. The effectiveness of EPA's implementation of the RFS will entirely determine the law's success.
- Congress should build on the foundation of the RFS by:
  - Adopting a low - carbon fuel standard that requires progressive reductions in the average greenhouse gas emissions per gallon of all transportation fuels sold, as California and Massachusetts are planning to do.
  - Passing comprehensive climate legislation built around a comprehensive mandatory cap on global warming pollution and an emission allowance trading system, such as that contained in the Climate Security Act reported by this Committee, with the value of allowances dedicated to public benefits.

INTRODUCTION

Thank you for the opportunity to share my views regarding the opportunities and challenges of implementing the Renewable Fuels Standard (RFS). My name is Nathanael Greene. I'm a senior policy analyst for the Natural Resources Defense Council (NRDC) and our director of renewable energy policy. NRDC is a national, nonprofit organization of scientists, lawyers and environmental specialists dedicated to protecting public health and the environment. Founded in 1970, NRDC has more than 1.2 million members and online activists nationwide, served from offices in New York, Washington, Los Angeles, San Francisco, Chicago, and Beijing.

Mr. Chairman, as you know, U.S. energy policy must address three major challenges: reducing America's dangerous dependence on oil, reducing global warming pollution, and providing affordable energy services that sustain a robust economy. Biofuels have the potential to contribute significantly to all three of these goals. Sustainably produced biomass feedstocks, processed efficiently and used in efficient vehicles can reduce our dependence on oil for transportation, reduce emissions of heat trapping carbon dioxide, and contribute significantly to a vibrant farm economy. Pursued without adequate guidelines such as those contained in current law, however, biofuels production carries grave risk to our lands, forests, water, wildlife, public health and climate. The potential for biofuels to be done right or wrong is reflected in recent headlines, which just a few months ago, regularly hailed biofuels as the solution to our oil addiction and now, roundly condemn biofuels in light of high food prices and recent studies that show how biofuel can increase global warming pollution and contribute to environmental degradation. While these concerns certainly should motivate greater efforts to get biofuels right, we need to be careful not to throw the baby out with the bathwater. We should go beyond all or nothing headlines and pursue a transition to biofuel strategies that realize the compatible objectives of replacing oil, expanding opportunities for biofuels producers, and securing both food supplies and a broadly sustainable future.

Replace Existing Biofuels Tax Credits and Tariffs with a Single Technology - Neutral, Performance Based Incentive

The existing biofuels tax credits and import tariffs are blunt, volume based policies that try to pick winners solely based on feedstocks. In doing so, these policies provide equal incentives to biofuels that cause negative environmental impacts as to biofuels that use the most beneficial practices and technologies. Similarly, while there is much debate about how big an impact today's biofuels have on food prices, these policies provide the same incentive to those biofuels that cause the most impact on food prices as those that cause none. For instance, the volumetric ethanol excise tax credit (VEETC) gives a fixed tax credit of $0.46 per gallon of corn ethanol and $1.02 per gallon of cellulosic ethanol regardless of how the ethanol was produced. Furthermore this tax credit is unavailable to butanol or biomass derived synthetic gasoline. Similarly the biodiesel blending tax credit is awarded on a per gallon basis regardless of whether the biodiesel is derived from palm oil grown in just cleared rainforests or waste grease diverted from a landfill. It's also not available to synthetic diesel. And our ethanol import tariff is similarly blunt.

With a mandate in place, and a biofuels industry with a track record of surpassing these mandate levels each year, it is time to reward performance and make sure we are not causing unintended consequences. Moreover, as the RFS ramps up the existing tax credits will become extremely expensive.

All these antiquated biofuels tax credits and tariffs should be replaced by a single technology neutral performance based incentive. Building off of the lifecycle GHG accounting protocol being developed for the RFS (discussed later), it would be relatively easy to link these incentives to improved GHG emissions, but I suggest that we go further. After all the RFS already starts us down the path towards biofuels with better GHG emissions and there are plenty of other ways that biofuels can help or hurt our environment and public health. I recommend that we use the tax credits and tariffs to encourage water efficiency, reduced water pollution, better soil management, enhanced wildlife management, and production of biofuels that do not displace food supplies, or better still produce more food and fuel. Developing accurate and workable accounting metrics for these impacts would be a non trivial challenge, but many of the tools we have developed to implement farm bill conservation programs could be used here.

While agriculture policy is the best place to deal with agriculture's environmental impacts on a broad basis, our biofuels policies should not be exacerbating these challenges. Perhaps most importantly, while the RFS will drive improvements to the performance of biofuels from new facilities, a new set of tax policies would drive improvements to our existing production as well.

Environmental Safeguards in the RFS

In contrast to the excise tax credit, the updated RFS represents an important step forward for biofuels policy in that it contains the basic minimum performance standards and incentives needed to promote biofuels that are part of the solution, rather than part of the problem. The challenge before us is to ensure that the law's critical safeguards are maintained and implemented aggressively and effectively.

I'd like to call your attention to four requirements under the updated RFS that were particularly far sighted of Congress to embrace and are critical to the law's success:

- Requiring conventional biofuels from all new facilities to achieve at least a 20 percent reduction in lifecycle greenhouse gas emissions compared to conventional gasoline and advanced biofuels to achieve at least a 50 percent reduction.
- Defining lifecycle greenhouse gas emissions to include the full cultivation, production, and combustion cycle of fuels and the emissions from both the direct and indirect land use change caused by this cycle. This critical step helps ensure biofuels produce real climate benefits and provides a direct disincentive to displacing food.
- Encouraging production of plentiful biofuels feedstocks—including woody-biomass—while ensuring the RFS mandate does not result in the loss of old-growth forest, native grasslands, "critically imperiled", "imperiled", "vulnerable" ecosystems pursuant to a State Natural

Heritage Program, the degradation of our federal forests [3], or conversion of natural forests on non-federal lands.

- Requiring the vast majority of new biofuels required under the law to be advanced biofuels derived from renewable cellulosic biomass with 60 percent reduction in lifecycle greenhouse gas emissions.

THE IMPORTANCE OF THE RFS'S MINIMUM LIFECYCLE GHG REQUIREMENTS

Section 201 of the RFS established minimum lifecycle GHG requirements for advanced and cellulosic biofuels. Section 202 established similar standards for conventional biofuel. To the best of my knowledge, these are the first lifecycle GHG standards established under any federal law. Under these standards, all renewable fuels from new facilities have to have lifecycle GHG emissions that are at least 20 percent lower than gasoline or diesel, depending on which they are replacing. In order to comply with the "advanced biofuels" definition, fuels need to have emissions that are at least 50 percent lower and to comply with the "cellulosic biofuels" definition, fuels need have emissions that are 60 percent lower.

This is the first time that biofuels policy in the US has required renewable fuels to proactively show an environmental benefit in return for benefiting from a government incentive program such as the RFS. Nowhere is the need for better performance more evident and urgent than when considering the global warming pollution impacts of biofuels. It is possible to produce ethanol derived from corn in ways that produce dramatically less lifecycle greenhouse gas emissions than the industry average for corn ethanol today. Conversely it is possible to produce ethanol from cellulosic feedstocks in a manner that produces far more lifecycle greenhouse gas emissions than gasoline (per BTU of delivered fuel). Unless our policies value, encourage and ultimately require biofuels to produce greenhouse gas reductions as the RFS has done for the first time, the market will provide whatever is cheapest and fastest. There is no reason to believe that the fuels the market would provide will be better than gasoline and plenty of reason to believe they would be worse.

THE RFS GETS THE DEFINITION OF LIFECYCLE GHG EMISSIONS RIGHT

Of course, the minimum lifecycle GHG standards for biofuels in the RFS would mean little without a good definition of lifecycle emissions. This is an area of the law where Congress showed particular foresight. Section 201(1)(H) of the RFS defines lifecycle GHG emissions as follows:

> '(H) LIFECYCLE GREENHOUSE GAS EMISSIONS.—The term 'lifecycle greenhouse gas emissions' means the aggregate quantity of greenhouse gas emissions (including direct emissions and significant indirect emissions such as significant emissions from land use changes), as determined by the Administrator, related to the full fuel lifecycle, including all stages offuel and feedstock production and distribution, from feedstock generation or extraction through the distribution and delivery and use of the finished fuel to the ultimate consumer, where the mass values for all greenhouse gases are adjusted to account for their relative global warming potential. [4] [Emphasis added.]

Less than two months after this definition became law, two articles that appeared in Science made it clear that the direct and indirect emissions associated with changes in land-use could dominate the lifecycle emissions of biofuels. The first article, "Land Clearing and the Biofuel Carbon Debt," addresses the direct greenhouse gas emissions from growing biofuel feedstocks on land recently converted from natural ecosystems to managed agriculture. [5] This article is authored by a team from the Nature Conservancy and the University of Minnesota including Dr. David Tilman. The second article, "Use of U.S. Croplands for Biofuels Increases Greenhouse Gases through Emissions from Land Use Change," addresses the emissions from land use change induced by the economic pressures when crops and land are diverted from food, feed, and fiber to fuels. [6] This article is authored by a team lead by Tim Searchinger from Princeton with other authors from the Woods Hole Research Center and Iowa State's CARD.

While there is little controversy over the notion that the emissions from lands converted specifically to produce biomass for renewable fuels should be accounted for in the lifecycle of those fuels, the first of these articles showed how large these emissions could be. For example, Fargione et al. estimate that burning a rainforest in Malaysia or Indonesia and converting it to a palm oil plantation and then using that oil to make biodiesel results in a release in carbon from the soil and the trees and other above ground biomass that takes 96 years to balance out through avoided emissions from petroleum based diesel. If the rainforest happens to have peat soil and that is drained and burned during the conversion, the breakeven period stretches out to 423 years. In other words, replacing petroleum based diesel with biodiesel from palm oil cultivated on land cleared from a peat soil rainforest will increase global warming pollution for the 423 years that the palm-oil biodiesel is produced.

The second article broke newer ground, pointing out that land conversion could be induced by biofuels if they increase the competition for arable land. Thus Searchinger et al. argue lifecycle accounting needs to look beyond just direct conversion of land for biofuels. Devoting an increased share of U.S. agricultural output to fuel production rather than food and livestock feed will result in increased demand for food and animal feed from sources abroad. If any significant portion of this additional food or feed is obtained by burning mature forests and converting them into pasture or cropland, the CO2 emissions from this land use change could greatly exceed the emission reductions from the use of biofuels.

The Argonne GREET model and most lifecycle analyses conducted to date have either ignored these land use related emissions or minimized them. These emissions, however, are driven by inevitable market economics when certain crops and types of land are use as biofuels feedstocks, and these economically mitigated emissions have the potential to negate all of the global warming benefits of poorly designed biofuels policies.

A recent letter in Science does a particularly good job of showing how complicated but important these indirect land use impacts can be. The letter explains how increased demand for corn to make ethanol is reducing domestic production of soy beans and thus driving up the production of soy beans in Brazil. The letter details how increased Brazilian soy farming leads directly and indirectly to clearing of Brazilian rainforests:

> Some Amazonian forests are directly cleared for soy farms. Farmers also purchase large expanses of cattle pasture for soy production, effectively pushing the ranchers farther into the Amazonian frontier or onto lands unsuitable for soy production. In addition, higher soy costs tend to raise global beef prices because soy - based livestock feeds become more expensive, creating an indirect incentive for forest conversion to pasture. Finally, the powerful Brazilian soy lobby is a key driving force behind initiatives to expand Amazonian highways and transportation networks in order to transport soybeans to market, and this is greatly increasing access to forests for ranchers, loggers, and land speculators. [Footnotes not included.] [7]

Not all biomass material leads to increased demand for new agricultural lands and not all lands brought into production are rainforests. Fargione et al. and Searchinger et al. acknowledge that global warming pollution reductions can be achieved from biofuels done right. Nevertheless, it is important to understand the scale of impact that greenhouse gas emissions from these indirect land use changes can have. Looking at a number of estimates, new very efficient corn ethanol refineries should be able to produce about 420 gallons of ethanol from an average acre of corn. Putting aside emissions from land use change, this ethanol would reduce greenhouse gas emissions by about 37 percent per gallon or about 2,500 pounds worth of CO2 per acre each year. Now, according to another article in Science, one acre of tropical rainforest burned and used to grow crops will release about 655,000 pounds worth of CO2 over 30 years or an average of nearly 22,000 pounds per year. [8] In other words, if the conversion of an acre of corn from food and feed to fuel resulted indirectly in the conversion of just one tenth of an acre of rainforest all the greenhouse gas emissions benefits of the ethanol would be wiped out for the first 30 years.

Of course, there are many more types of land being converted to agriculture than just rainforests. And the marginal impact of land use changes here in the United States on land use in the rest of the world is extremely hard to predict with economic equilibriums and agricultural and trade policies all interacting in complex ways. But to ignore these indirect emissions is to assume they are zero, which could easily lead to the government subsidization of fuels that are worse for global warming than gasoline or diesel.

While these two articles have already stirred a lot of debate about the specific amounts of carbon released from different land types, the amounts of different lands being cleared, and the exact economics driven by growth in biofuels production, three conclusions are clear now: 1) absent the GHG standards in the RFS and the carefully crafted definition of lifecycle emissions, these two dynamics make it very likely that most biofuels would be responsible for greenhouse gas emissions significantly higher than gasoline or diesel; 2) the fundamental dynamics addressed by these two articles (direct land use emissions and economically induced land use emissions) are driven by the fundamentals of soil science and the laws of supply and demand; and 3) the importance of implementing the minimum GHG emissions standards and land use safeguards in the RFS aggressively and effectively is clearer than ever. I return to this last point later in my testimony.

Finally, it is important to emphasize the direct relationship between land use change, GHG emissions, and food price disruptions. The same economic factors that link biofuels to indirect emissions from land use change also link biofuels to food price increases. It is the diversion of food and feed crops and land that could produce food crops that drive up the competition for land and subsequently land use change and drive up the competition for food and feed and subsequently their prices. Requiring land use change to be included in the lifecycle definition provides a direct disincentive to divert food crops to fuel, and an incentive to produce biofuels from feedstocks that do not disrupt food production—such as crop residues, biomass grown on degraded or marginal land, or feedstocks that produce more food than current crops and fuel from the same acre of land.

The RFS Includes Critical Land and Wildlife Safeguards

In addition to the minimum GHG standards, the RFS includes a definition of renewable biomass that provides essential safeguards for wildlife, native grasslands, old-growth, natural forests, and federal forests. At the same time, it is broadly inclusive of the kind of material that typically provides the biggest sources of biomass, assuring diverse opportunities for landowner participation and a wide diversity of feedstocks.

Eligible Biomass

The renewable biomass definition includes:

- All crops and crop residue from current agriculture land and non-forested, fallow land
- All crops and crop residue from any non-forested land cleared prior to the enactment of the Energy Independence and Security Act of 2007 (EISA07), including newly established tree plantations [9]
- All trees and logging residue from non-federal tree plantations, excluding those converted from natural forests after passage of EISA07 (See below)
- "Slash and pre-commercial thinnings" from non-federal natural forests, which, importantly, constitute the lion's share of woody-
- biomass from natural working forests that are expected to be economically viable options for biofuels, while keeping forests from being converted

- All material removed from the immediate vicinity of homes and communities at risk from wildfire, on federal and non-federal lands
- Animal waste and animal byproducts
- Waste material, including separated yard waste, food waste, and cooking and trap grease

Protecting Wildlife

The definition of renewable biomass ensures the RFS does not encourage biomass harvesting from sensitive wildlife habitat. The RFS employs the State Natural Heritage programs to identify critically imperiled, imperiled and vulnerable wildlife habitat. The Natural Heritage programs are readily accessible, widely recognized, and embraced by all 50 states. They are the leading sources on the precise locations and conditions of rare and threatened species and ecological communities found within each state. These databases and ranking systems are used effectively for forest management and in partnership with many forest-product industry leaders.

The ecosystems identified by the RFS as off-limits are home to our most rare, threatened, and imperiled wildlife. While tree plantations and young forests are increasing in parts of the United States, older forests that provide critical wildlife habitat and store tremendous amounts of carbon are disappearing faster than they are being regrown, both nationally and globally, and loss of native habitat is the greatest threat to biodiversity here and abroad. Animals are currently going extinct at a rate nearly 1,000 times higher than they have historically, and under current trends that may increase to 10,000 times over the next century. [10] Moreover, as global warming escalates, wildlife is increasingly threatened by loss of safe harbors and migration routes, making habitat protection even more important. The RFS safeguards ensure that the law's new demand for feedstocks does not translate into irreversible loss of these at risk habitats.

Native Grasslands and Old-Growth Forest

The RFS safeguards also protect against the use of biomass harvested from native grasslands and old-growth and late successional forest. Native grasslands represent one of the most threatened ecosystems in the world. Less than 4 percent of our country's original native prairies exist today. These imperiled ecosystems represent a last remnant of our natural heritage and provide invaluable habitat for migrating birds and other endangered species. Similarly, our remaining old-growth trees constitute a rare and vulnerable ecosystem type that provides unique wildlife habitat, water filtration, and ecosystem resiliency. Nationally, old-growth forests are severely diminished. In the lower 48 states, old growth forest makes up just 2 percent of the remaining forest. [11] As we struggle to maintain and restore these ancient forests, it is imperative that federal policy not further their endangerment.

Conversion of Natural Forests

Loss of forests is one of the greatest threats to biodiversity worldwide and a major contributor to global warming. [12] Natural forests are under severe threat from unsustainable logging practices, global warming, and real estate development. While deforestation is the most dramatic example of this growing crisis, equally critical is the conversion of natural forests to single species tree plantations. Plantations may look like "forests," but they are biological deserts when compared to the natural forests that they replace—lacking the diversity of species, structure, and ecological functions that make natural forests so important.

A potent example of conversion's sweeping impacts can be found in the forests of the Southern United States which contain some of the most biologically rich forests in North America, housing an abundance of plant and animal diversity that exist nowhere else in the world. Unfortunately, these unique forests are under increasing pressure from the wood products industry as well as urban sprawl and development. Pine tree farms have been displacing natural forests for the past 50 years and now occupy 32 million acres (15 percent) of the current Southern "forest." [13] Seventy-five percent of the pine plantations established in the last two decades were carved out at the expense of natural forests. Moreover, 40 percent of the region's native pine forests have already been converted to single-species plantations, eliminating the rich diversity that the area is known for.

The RFS definition of renewable biomass does not by any means exclude woody biomass, but does ensure that federal policy is not making this bad situation worse. The RFS renewable biomass definition includes all biomass from existing tree plantations, new tree plantations established on previously cleared non- forested lands, and "slash and pre-commercial thinnings" from natural forests. In concert, these provisions allow woody-biomass to contribute to biofuels, while protecting against the clearing of forests or the conversion of natural forests to monoculture tree plantations, thus losing their natural ecosystem functions. It is important to emphasize that we believe the term "slash and precommercial thinning" should be interpreted with substantial flexibility - allowing the use of all harvest byproducts, as well as small and low-value trees from natural forests, as long as the forest is naturally regenerated after harvest as opposed to converted into a tree plantation or other crop.

Sustainable forestry practices that identify and protect high conservation values such as old-growth or late successional forest and specific wildlife habitat, and avoid conversion, are well established. These practices allow natural forests to remain working forests, without sacrificing critical wildlife habitat and other important environmental values. For example, Forest Stewardship Council certification, a global standard used in the forest products industry, incorporates these considerations.

Federal Forests

Our federal forests represent unique reservoirs of biologic diversity, genetic diversity, significant carbon stores, and many other ecological services, and stand to play a critical role in the face of global warming's growing impacts, including loss of biodiversity, decreased ecosystem resilience, and the spread of invasive species. [14] It is therefore becoming commensurately more important that our federal forest resources are managed and preserved for their numerous non commodity values and that we assiduously avoid policies that would impose additional pressures on these already stressed, and increasingly crucial, public resources.

In this context, proposals like those contained in H.R. 5236 and S. 2558 to use "preventative thinnings" from national forests as a biofuels source make little economic or ecologic sense. First, it is important to understand that preventative thinning—the removal of forest biomass including anything from small brush to large trees to address forest health—is essentially logging and thus not devoid of ecological impacts, such as soil compaction, spread of invasive species, hydrologic disruption, and in the case of associated road building, increased fire risk due to lost resiliency and increased human traffic. [15]

The argument for the production of biofuels from national forest preventative thinnings hinges on three basic assumptions, all three of which would have to be valid for the proposition to be worth the impacts and risks of logging: first, preventative thinnings based biofuels do not negatively impact global warming; second, preventative thinnings will safely and sustainably produce a meaningful volume of biofuels; and third, biomass removal is beneficial to addressing wildfire. Unfortunately there is uncertainty and debate around each of these assumptions.

The GHG benefit of preventative thinnings for biofuels is highly uncertain. As noted above, preventative thinning represents the removal of biomass—or stored carbon—through mechanical harvest. For preventative thinnings to make sense from a GHG perspective, the fuel produced would have to be "better" than the lost carbon storage, including soil carbon, the emissions resulting from the removal, transportation, and processing of the biomass, and the burning of the final fuel. It is also important to note that fire risk reduction thinning, even where appropriate (see below), is successful only to the extent that occasional intense burns are replaced by cooler burns that occur perhaps 20 to 25 times more often. While ecotype specific data are still not available, on the face of it, the much more frequent burns are likely, if anything, to result in greater emissions.

Even if preventative thinning were ecologically necessary, most scenarios indicate a limited supply of material within economic haul distances, making biofuels from preventive thinning at best a drop in the overall bucket. [16] Preventative thinnings are single entry activities pursued for restoration purposes and do not provide a renewable resource from any given location. Thus they are severely constrained by the energy and economic costs of transporting biomass from individual sites to central processing facilities. Incenting the establishment of a whole industry in order to supply a negligible volume of fuel from a time-limited supply of any arguably legitimate feedstock presents likely negative outcomes, including either a boom-bust cycle, or future pressure to shift to an unsustainable scale of extraction. This is particularly unappealing considering there are other, proven, and more readily scalable uses for harvest and preventative thinning byproducts where it is economic to remove them from the woods, such as community heat and electricity production and manufactured products. These factors are particularly important when considering utilization of slash and byproducts from sources other than preventive thinning, including any backlog like slash piles. While this material may be available for the short term, it would soon be exhausted, representing a nonrenewable supply far more appropriate for more scalable uses than biofuels. [17]

Finally, while intuitively appealing, the empirical evidence is mixed at best on whether backcountry logging and preventative thinning effectively reduces fire risk [18], and indicates it may in fact increase the chances of uncharacteristic fire. [19] Furthermore, it is a mistake to conceive of national forests as uniformly overgrown thickets in need of preventative thinning to restore prior forest structure and fire regimes. While evidence suggests some lower elevation, dry forests could benefit from restoration treatments, many other sites across the country, including lodgepole pine, spruce-fir forests, subalpine forests, piñon juniper, mixed conifer systems, and ponderosa pine, are adapted to intense, stand-replacing fires, and in these dense stands preventative thinning is contraindicated. [20] The empirical evidence on both the efficacy and necessity of preventative thinning suggests it is still experimental, poses significant risks, is constrained to limited areas at best, and therefore should be pursued only on an investigational basis.

In sum, none of three underlying assumptions related to producing biofuels from preventative thinnings reflect the best available science or pragmatic, on the ground scenarios. To contribute a negligible amount of fuel, we would have to risk further degraded forests, exacerbating fire risk, reducing carbon storage, increasing GHG emissions, and establishing an unsustainable industrial demand for continued commercial exploitation of vital public resources.

THE RFS CORRECTLY FOCUSES PRIMARILY ON BIOFUELS FROM RENEWABLE CELLULOSIC BIOMASS

While the RFS requires 36 billion gallons of biofuels by 2022, only 28.5 of this is additional to the previous RFS and only about 24 is in addition to what the market would have almost certainly provided on its own. The new RFS requires that at least 22 billion gallons of the 36 billion total be "advanced biofuels," which are basically defined as not being ethanol from corn. As mentioned earlier, these advanced biofuels must provide at least a 50 percent reduction in global warming pollution compared to petroleum fuels. Of the advanced biofuels, at least 16 billion must be from cellulosic feedstocks and at least 1 billion must serve as an alternative to petroleum diesel. The advanced biofuels from cellulosic feedstocks must provide at least a 60 percent reduction in GHG emissions.

Much has been written and said about the promise of advanced, second generation biofuels technologies. These technologies do appear poised to greatly increase the amount of biofuels we can produce and make it easier to produce them in a sustainable way. It is critical to realize, however, that these technologies will not be available overnight and just because we can produce biofuels sustainably does not mean that we will.

When I first started looking at biofuels in 2002, all of the cutting edge expertise was in academia and the national energy labs. You could talk to these experts, and they would tell you where the technology stood. Over the last 2 years, however, all of the cutting edge research has moved into the private sector and is proprietary. So while it s now much harder to know where things stand, we know that a lot of investor dollars are being bet on near term commercialization. The research is being driven by venture capitalists and private investors.

Combine these developments with the very impressive number of projects proposed in response to recent government solicitations, and it s hard not to believe that things are moving along quickly. Within the past year, New York issued a solicitation for two pilot cellulosic biofuels projects and DOE issued a solicitation for six small commercial scale cellulosic projects and seven more pilot scale cellulosic projects. All of these solicitations required significant private sector investment and a number of major market players responded. Cellulosic biofuels projects announced in recent months include a new pilot cellulosic plant in Nebraska that will be built by Abengoa, a plant using switchgrass as a feedstock that will be constructed in Tennessee by Mascoma and a commercial line of cellulose processing enzymes by Genencor. International developments include a recent announcement by Royal Nedalco in the Netherlands that it will skip the pilot scale and go straight to building a small commercial scale 50 million gallon a year cellulosic plant. There are also advances being made in radically different technologies including the use of microorganisms in existing ethanol facilities to produce fuels similar to gasoline such as biobutanol, bacterial and catalytic conversion of biomass into renewable diesel and gasoline, and the use of algae to make a synthetic diesel fuel.

It is my understanding, however, that none of these projects will come on line until late next year at the earliest. Assuming a few of them perform very well, they could be expanded, but it is really the second generation plant that investors will consider a potential cookie cutter model. Being optimistic, assume that we go into 2013 with three different technologies that can compete with corn ethanol or gasoline, each with an operating second generation plant of about 50 million gallon per year capacity for a total of 150 million gallons annual capacity. Even if the technologies are so promising that orders for more plants are actually placed in 2012, how fast will capital and engineering capacity flow into the sector? How long will siting and permitting lead times be? One billion gallons of capacity by 2016 seems eminently achievable to me assuming we have at least one clear success on line by 2010. Three billion would be absolutely fantastic. Such a result would require that by 2013 the cellulosic industry grows as fast as the corn ethanol industry grew from middle of 2006 to middle of 2007.

In comparison, the RFS requires 1 billion gallons of cellulosic biofuels by 2013 and 3 billion by 2015. These levels are extremely aggressive, which is why EISA07 explicitly allows EPA to reduce the levels if the forecasted production levels are below the forecasted production levels. This effectively provides producers of all the capacity that is likely to come on line between now and 2015 with a guaranteed market.

The ability to convert cellulose into fuels opens up the possibility of using new feedstocks such as cellulosic crops—including switchgrass—that use significantly less chemical inputs and water, agricultural residues and organic waste. However, as we discussed earlier, it is also possible to cultivate and harvest cellulosic biomass in extremely destructive and carbon intensive ways. One of the easiest ways to do cellulosic biofuels wrong is by harvesting feedstocks from inappropriate areas such as our public forests, old growth forests, or other imperiled and fragile ecosystems. Federal policies should not incentivize the use of such feedstocks. To the contrary, our policies should encourage the best practices and the best choices in feedstocks through technology and crop neutral, performance based standards and incentives. Environmental safeguards and performance standards are necessary to ensure that federal policy promotes the best production standards for biofuels and then let the market decide which feedstocks can compete.

Given the emissions from land use change associated with food crops and arable land, many wonder if there is enough truly low carbon biomass available to make biofuels that comply with the RFS's minimum lifecycle GHG standards. There are two reasons to be very cautious about making any forecasts about the supply of low-carbon biomass. First EPA has not written its regulations, so it is impossible to know exactly which types of biomass will be able to comply. Second, the market has never faced the combination of commodity food prices and regulation driven prices. This combination makes any feedstock that reduces food or fiber production highly unattractive, economically—even if a farmer or a forester was willing to forego the market value of food or fiber, the greenhouse gas emissions that would be associated with displacing those products would probably make the resulting biofuels non compliant with the RFS standards. This means that farmers and foresters will have a strong incentive to innovate and to find ways that produce biomass without reducing food and fiber production or, better yet, produce more of both. We can articulate examples of what this type of innovation might look like, but the reality is that there inevitably will be innovations that we can't currently imagine and it's impossible to predict which innovations will triumph.

Nevertheless, a preliminary and partial assessment of potential sustainable and truly low-carbon feedstocks does suggest that we have more than enough cellulosic biomass to meet the RFS's 16 billion gallon requirement for 2022. For example, cover crops have long been recommended as a conservation practice to reduce soil erosion and fertilizer runoff, but only about 10 million acres are cover cropped each year and most of the biomass that is produced is plowed under. The right price for biomass could lead to a great expansion in the number of acres and the development of cover crops that can yield 2 to 4 tons per acre. The challenge to regulation is to encourage this expansion to happen in a way that maintains as much of the conservation value as possible and minimizes any reduction in yield of the primary crop. The US has nearly 400 million acres in annual row crops. If just 25 percent of this could be cover cropped with a yield of just 2 tons per acre on average, that would produce about 200 million tons per year. At a conversion rate of 80 to 100 gallons of ethanol per ton, this would supply between 16 and 20 billion gallons—equal to or greater than the RFS requirement—and potentially help greatly reduce soil erosion and water pollution.

In Growing Energy, my coauthors and I wrote about another example. [21] Farmers currently grow about 70 million acres of soybeans primarily for feed. Switchgrass provides equal or great yields of feed protein per acre as soybeans and on average 5 tons per acre of biomass today. Switchgrass is also a perennial that increases soil carbon, needs less water, and can provide a better wildlife habitat. If 40 million acres of soy beans were converted to switchgrass, this would also provide about 200 million tons per year—equal to or greater than the amount needed to comply with the RFS — and improving soil quality and wildlife habitat.

Add to these examples, agricultural residues, forest residues, construction and demolition wood waste, and other clean biomass waste and it becomes clear that the supply of sustainable, low-carbon biomass need not be a limiting factor for advanced biofuels for a long time to come. However it also becomes clear that we have a lot to learn about how to cultivate and manage these types of feedstocks. In fact, especially in light of the private sector's major push to develop conversion technologies, the most important role for federal RandD dollars right now is helping to develop innovative, broadly sustainable, and truly low carbon biomass supplies.

The RFS's Environmental Safeguards Can and Must be Effectively Implemented by EPA

While Congress deserves much credit for carefully crafting the standards, safeguards, and study provisions of the RFS, none of these will amount to a hill of beans unless they are aggressively and effectively implemented. Under the RFS, EPA is directed to promulgate regulations to implement these GHG performance standards and the environmental safeguards by the end of 2008. EPA's task is complex, but hardly insurmountable. There are essentially four parts of the legislative requirements that EPA's regulations must knit together. First, they must provide a methodology for assessing the indirect emissions. This is likely to start off as something relatively simple such as a look up table with different feedstocks and land types having different values. It should include, or evolve to include, a method for certifying a unique value to accommodate innovations.

Next the feedstock must be certified in compliance with the definition of renewable biomass in EISA07 and the emissions associated directly with the cultivation and harvesting of the feedstock must be calculated. Again there is likely to be a relatively simple set of default values for most feedstocks with the option for customized values. Importantly though, this stage does require site specific certification to ensure compliance with the definition of renewable biomass. There are many examples of certification and information tracking that EPA can use as a model for its regulations. Third party certification, in which certifiers are trained and "accredited" to evaluate and verify management standards has been widely adopted by the forest products and organic agriculture industries. The third-party certifiers, who do site visits and verify claims, include major auditing firms that we have all heard of such as PricewaterhouseCoopers. Similarly, The National Organic Program (NOP), administered by the Agricultural Marketing Service of the USDA, is carried out by state, private, and foreign organizations or individuals that AMS accredits to become "certifying agents." These agents are responsible for receiving producer/handler applications, certifying them, and verifying that they are complying with the NOP standards.

Third the emissions of the conversion facility must be included in the lifecycle calculations. These are much simpler in many ways because they are point' source emissions and overwhelmingly a function of the type of fossil fuels used and the equipment in which the fossil fuels are used.

Finally, once the feedstock has been certified as "renewable biomass" and the lifecycle GHG emissions have been calculated, this information must be attached to each gallon or batch of gallons of fuel and tracked through the wholesale market and the oil companies must be held accountable for compliance with the RFS gallon requirements. Fortunately, EPA's existing "Renewable Identification Number" system is in place and easily amended to include this additional information.

In implementing the RFS, EPA should strive to meet the following three criteria to ensure the standards are applied in the most technology neutral, performance based, and fair manner:

EPA should Use on the Best Available Science and Modeling to Determine Lifecycle GHG Emissions and when Confronted by Uncertainty not Assume there are no Impacts

Perhaps the most complicated part of this is developing the accounting protocol to measure and certify the lifecycle greenhouse gas emissions of different renewable fuels. Fortunately, EPA has a head start in this effort. Early in 2007, President Bush directed EPA, in coordination with other federal agencies, to promulgate regulations to reduce US gasoline use by 20 percent within 10 years and to do so in a way that complied with the federal court ruling that CO2 is a pollutant. Before the passage of the EISA07, EPA was on track to issue a notice of proposed rulemaking to implement the so called 20 in10 executive order around the end of 2007. As part of these draft rules, EPA had done significant work developing a lifecycle accounting methodology.

While there is every indication that EPA is bringing together the best science and, for the indirect emissions, the best economic modeling, there will no doubt be some that will argue that the agency should abandon this and either permanently or temporarily ignore certain sources of emissions. Some will argue that the indirect emissions, in particular, are too uncertain and the modeling too new and therefore these emissions should not be included. Others will argue that either Congress did not mean for EPA to include international emissions in the lifecycle of domestically produced fuels or that EPA does not have the authority to include international emissions. EPA should reject these arguments.

EPA regularly has to assign values to impacts that are uncertain and as discussed earlier, the scale of the land use emissions can be so large that to ignore it or assume it is zero would be to turn a blind eye to what will be the largest source of emissions in many cases. Therefore the uncertainty must be weighed against the potential scale of impacts. Assuming the indirect emissions are zero clearly risks sending grievously wrong signals to the market. Some might look at this uncertainty and argue that instead of zero, we should assign an infinite value and simply stop implementing the RFS. However, this ignores the important options for doing biofuels in the right way. Only by using the best science and modeling to assign values to as many different options as possible can EPA give guidance to the market while allowing innovation.

Excluding the international component of indirect emissions would be arbitrary and would effectively exclude a large portion of these emissions. In the definition of lifecycle GHG emissions (see quote above), Congress directed the EPA to include the "aggregate" emissions "related to the full fuel lifecycle." The plain text reading of this includes international sources especially in light of the fact that Congress specifically included and excluded domestic and international items from other programs under EISA07.

Nor does including economically induced, international emissions violate EPA's authority. To be clear, NRDC does not advocate, and believes that EPA does not contemplate, direct regulation of international land use change. EPA does, however, have the authority to recognize indirect emissions induced through international economic markets in its analysis of lifecycle greenhouse gas emissions. Furthermore, agencies regularly recognize the economic and market impacts of their regulations, and may modify their regulations according to these considerations.

EPA should Regularly Update RFS Regulations

While the various elements of science and modeling that EPA is likely to use in developing default values for the lifecycle GHG emissions of different feedstocks have been widely used and studied, there can be no doubt that the combination needed to implement the RFS is novel. Furthermore, in trying to get regulations in place in a timely manner, EPA is necessarily going to be limited by the amount of customization that it can allow in the first generation of regulations. Overtime, as innovations proliferate, rigidness in the first generation of EPA's rules could act as a significant barrier to new paths. Thus to incorporate the best new science and modeling and expand the options for new and different biofuels' pathways, EPA should regularly update its regulations. These updates should draw on external science advisors such as the National Academy of Sciences and on the studies that EPA must carry out under EISA07. I recommend that an initial update be planned for roughly 3 years after the first regulations are promulgated.

EPA Should Require Site Specific, Audited Information as the Basis of Certifying Lifecycle GHG Emissions and Compliance with the Definition of "Renewable Biomass"

Much of the information that EPA, or accredited certifiers, will need to determine the lifecycle GHG emissions of different biofuels and compliance with the definition of "renewable biomass" can only be gathered on the farm, in the forest, or at the biofuel refinery. As discussed earlier, there are a variety of working models for auditing and certification of this type of information. Farm Bill conservation programs may provide further examples of ways to collect and verify site specific information. While we recognize that this will pose an implementation challenge to EPA and a new cost of doing business on the biofuels industry, we believe that it is the only way to ensure the RFS GHG standards and environmental safeguards are effectively implemented. Congress must make sure EPA is fully funded to both develop the implementing regulations and then carry out the enforcement and studies. As for the cost to industry of compliance, we believe it is a small price for the industry to pay compared to the benefits it gets from the federal mandate for biofuels.

Our discussions with staff within EPA give us confidence that the agency is make real progress towards a workable, science based set of regulations. Under EISA07, technically EPA should promulgate these regulations by the end of this calendar year. Given the genuine complexity of the issues that have to be addressed, this timing seems unrealistic, but given the progress that we see EPA making, we're confident that they're on track to finish the rules within a reasonable period. Nevertheless, we encourage Congress to monitor their progress closely to ensure that science rather than politics drives the resulting regulations.

The effectiveness of EPA's implementation of the RFS will entirely determine the law's success.

CONGRESS SHOULD BUILD ON THE FOUNDATION LAID BY THE RFS

I recommend the following steps to build on the foundation of the RFS:

Congress Should Adopt a Low Carbon Fuel Standard Like California and Massachusetts are Doing

Adopting a low carbon fuel standard (LCFS) that requires progressive reductions in the average greenhouse gas emissions per gallon of all transportation fuels sold, as California and Massachusetts are planning to do. The LCFS is a technology neutral, performance based approach to reducing the greenhouse gas emissions from transportation energy. This would be an important improvement over the technology specific, volume incentives and mandates that until recently dominated US biofuels policies.

The way a LCFS works is that the full lifecycle GHG emissions from the fuels each oil company is selling are added up and divided by all the energy in that fuel. This becomes the company s average fuel carbon intensity. Overtime under the LCFS, the oil companies have to reduce this average carbon intensity by mixing in sources of transportation energy with lower lifecycle GHG emissions. In California, which was the first to move towards a LCFS and is now in the process of developing the regulations, the goal of the LCFS is to require a 10 percent reduction in carbon intensity by 2020. In other words, a company could replace all of their current fuel with an alternative that has 10 percent lower lifecycle GHG emissions, or half with a 20 percent lower alternative, and so on.

The LCFS rewards the sources of energy that have the lowest lifecycle GHG emissions. Just as importantly, it penalizes high carbon fuels such as liquid coal.

This is in contrast to the original RFS, which was simply a volume mandate that almost totally ignored how the biofuels were produced. Our current tax credits for ethanol and biodiesel and our import tariff on ethanol are similarly blunt, ignoring the impacts or benefits of the fuels lifecycle. While the current RFS is the first step towards setting performance based requirements, it is still a volume mandate for a specific set of fuels and these standards are floors. Electricity and natural gas can't be used to comply and there's no incentive for producing biofuels that perform better than minimum standards.

Congress Should Pass Comprehensive Climate Legislation Adopting a Carbon Cap and Trade System

It is much harder to get biofuels right in the context of a broader economy where greenhouse gas emissions are not regulated. In order to meaningfully level the field between oil and renewable fuels and encourage the economy wide changes in practices needed to drive a sustainable transportation sector, we need comprehensive approach to global warming. In addition to a low carbon fuel standard this should include an economy wide carbon cap and trade system.

Senate bill S.2191, the Lieberman Warner bill, includes both, Congress should pass this bill and the President should sign it as soon as possible.

Conclusion

Renewable fuels hold great promise as a tool for reducing global warming pollution, breaking our dangerous oil addiction, and revitalizing rural economies, as long as appropriate standards and incentives are used to shape the nascent bioenergy industry to provide these benefits in a sound and truly sustainable fashion. Congress deserves credit for the foresight it showed in starting to build these standards and safeguards into the new RFS. We should build on this foundation by making over the rest of our biofuels policies to be technology neutral and performance based. I look forward to working with the EPA to implement the RFS and with the Committee to continue to improve our biofuels policies.

REFERENCE

[1] 1Biomass obtained from the immediate vicinity of buildings and other areas regularly occupied by people, or of public infrastructure, at risk from wildfire is excepted from these restrictions, on both federal and non-federal lands.

[2] Section 211(v)(2)(A) of the Clean Air Act (42 U.S.C. 7545) as amended by Section 209 of EISA07.

[3] Biomass obtained from the immediate vicinity of buildings and other areas regularly occupied by people, or of public infrastructure, at risk from wildfire is exempted from these restrictions, on both federal and non-federal lands.

[4] "Energy Independence and Security Act of 2007," Title II, Section 201(1)(H), signed into law on December 17, 2007.

[5] Fargione, J., et al., "Land Clearing and the Biofuel Carbon Debt," *Science* [DOI: 10.1126/science.1152747] February 7, 2008.

[6] Searchinger, T., et al., "Use of U.S. Croplands for Biofuels Increases Greenhouse Gases Through Emissions from Land Use Change," *Science* [DOI: 10.1126/science.1151861] February 7, 2008.

[7] Laurance, W.F., "Switch to Corn Promotes Amazon Deforestation," *Science* 318, no. 5841 (December 14, 2007): 1721, DOI: 10.1126/science.318.5857.

[8] Renton Righelato and Dominick V. Spracklen, "ENVIRONMENT: Carbon Mitigation by Biofuels or by Saving and Restoring Forests?," *Science* 317, no. 5840 (August 17, 2007): 902, doi:10.1126/science.1141361.

[9] While I recognize that the term'plantation' carries negative historical connotations, it is used throughout my testimony because'tree plantation' it is a technical term distinct from'tree farm'. 'Tree plantation' is also the term used in the Renewable Biomass definition legislative text.

[10] "Environmental Science and Engineering for the Twenty First Century: The Role of the National Science Foundation," National Science Foundation, February 2000; Peter Raven, "Plants in Peril: What Should We Do?" Missouri Botanical Garden, 1999.

[11] Palmer, T., The Heart of America: Out Landscape, Our Future, Island Press, 1999.

[12] Intergovemental Panel on Climate Change, Climate Change 2007: Synthesis Report Summary for Policymakers, pg. 5. Available at http://www.ipcc.ch/pdf/assessment-report/ar4/syr/ar4_syr_spm.pdf

[13] See USFS SFRA 2001 Summary–Section 3.2.2

[14] See, for example, Lovejoy, Thomas, Climate Change and Biodiversity, Yale University Press, August 2006.

[15] The literature on the ecologic impacts of logging and road building is extensive. For a collection of independently reviewed material, see http://www.nrdc.org/land/forests/roads/eotrinx.asp. See also USDA. "Roadless Area Conservation Final Environmental Impact Statement." US Forest Service. Vol. 1. (November, 2000). pp. 3 116. Eastman, J. C., et al. "Roadless Areas and Forest Fires in the Western United States." American Geographical Union Spring Meeting. (May 29, 2002). Pyne, S. J. Tending Fire: Coping with America's Wildland Fires, Island Press, 2004, p. 208.

[16] For example, the DOE "Billion Ton Study" available at http://www1.eere.energy.gov/biomass/pdfs/finalbilliontonvisionreport2.pdf reports only 11.7 million dry tons of biomass available from national forest preventative thinnings. Even this estimate may be optimistic due to economic costs, haul distances, and serious questions regarding ecological impacts.

[17] See DOE "Billion Ton Study "estimate of only 1.5 million dry tons of national forest logging residue, under future optimistic conditions.

[18] See, Martinson, E. J. and P. N. Omi. 2003. Performance of Fuel Treatments Subjected to Wildfires, in Omi, P. N.; Joyce, L. A., technical editors. Fire, fuel treatments, and ecological restoration: Conference proceedings; 2002 16-18 April; Fort Collins, CO. Proceedings RMRS-P-29. Fort Collins, CO: U.S. Forest Service, Rocky Mountain Research Station. pp. 7-8. See also Carey, H. and M. Schumann. 2003. "Modifying Wildfire Behavior-The Effectiveness of Fuel Treatments." The Forest Trust. p. 16. Available at www.theforesttrust.org/images/swcenter/pdf/WorkingPaper2.pdf. p. 15 ("The proposal that commercial logging can reduce the incidence of canopy fire appears completely untested in the scientific literature"). See also Cram, D.S., T.T. Baker, and J.C. Boren. 2006. Wildland Fire Effects in Silviculturally Treated vs. Untreated Stands of New Mexico and Arizona. Research Paper RMRSRP-55. Fort Collins, CO. U.S. Forest Service, Rocky Mountain Research Station. p. 1. ("information comparing fire behavior and

fire effects on treated versus untreated forest stands following wildland fire remains largely anecdotal.")

[19] Martinson and Omi, supra note 1. p. 7. U.S. Forest Service. 2000a. Final Environmental Impact Statement for the Roadless Area Conservation Rule ("FEIS"), volume 1. Online at: http:/www.roadless.fs.fed.us/documents/feis. p. 3-110. Collins, B.M. et al. 2007. Spatial patterns of large natural fires in Sierra Nevada wilderness areas. Landscape Ecology 22:545-557. p. 554. Whitehead, R.J. et al. 2006. Effect of a Spaced Thinning in Mature Lodgepole Pine on Within- stand Microclimate and Fine Fuel Moisture Content, in Andrews, P. L. and B.W. Butler, comps., Fuels Management-How to Measure Success: Conference Proceedings. 28-30 March 2006; Portland, OR. Proceedings RMRS-P-41. Fort Collins, CO: U.S. Forest Service, Rocky Mountain Research Station. Online at http://www.fs.fed.us/rm/pubs/rmrs_p041/rmrs_p041_523_536.pdf. p. 529. Keeley, J.E., D. Lubin, and C.J. Fotheringham. 2003. Fire and grazing impacts on plant diversity and alien plant invasions in the southern Sierra Nevada. Ecological applications 13:1355-1374. p. 1370. FEIS, supra this note, Fuel Management and Fire Suppression Specialist's Report. Online at: http://www.roadless.fs.fed.us/documents/feis/specrep/xfire_spec_rpt.pdf. p. 21 ("Fahnstock's (1968) study of precommercial thinning found that timber stands thinned to a 12 feet by 12 feet spacing commonly produced fuels that 'rate high in rate of spread and resistance to control for at least 5 years after cutting, so that it would burn with relatively high intensity;'" "When precommercial thinning was used in lodgepole pine stands, Alexander and Yancik (1977) reported that a fire's rate of spread increased 3.5 times and that the fire's intensity increased 3 times"); id. At 23 ("Countryman (1955) found that 'opening up' a forest through logging changed the 'fire climate so that fires start more easily, spread faster, and burn hotter").

[20] See Christensen, N, et al. 2002. Letter to President George W. Bush http://docs.nrdc.org/land/lan07062801g.pdf; Romme, W. et al. 2006. Recent Forest Insect Outbreaks and Fire Risk in Colorado Forests: A Brief Synthesis of Relevant Research. Colorado State University, Fort Collins, CO. Online at http://www.cfri.colostate.edu/docs/cfriinsect.pdf. Schoennagel, T., T.T. Veblen, and W.H. Romme. 2004. The interaction of fire, fuels and climate across Rocky Mountain forests. BioScience 54: 661-676. p. 666. Romme, W., et al. 2003. Ancient Piñon-Juniper Forests of Mesa Verde and the West: A Cautionary Note for Forest Restoration Programs, in Omi, P. N.; Joyce, L. A., technical editors. Fire, fuel treatments, and ecological restoration: Conference proceedings; 2002 16-18

April; Fort Collins, CO. Proceedings RMRS-P-29. Fort Collins, CO: U.S. Forest Service, Rocky Mountain Research Station. Baker, W.L. and D.S. Ehle. 2003. Uncertainty in Fire History and Restoration of Ponderosa Pine Forests in the Western United States, in Omi, P. N.; Joyce, L. A., technical editors. Fire, fuel treatments, and ecological restoration: Conference proceedings; 2002 16-18 April; Fort Collins, CO. Proceedings RMRS-P-29. Fort Collins, CO: U.S. Forest Service, Rocky Mountain Research Station. p. 330.

[21] Greene, N. et al., "Growing Energy: How Biofuels Can Help End America's Oil Dependence," Natural Resources Defense Council, December 2004. http://www.nrdc.org/air/energy/biofuels/biofuels.pdf

In: Renewable Fuel Standard Issues
Editor: Daniel T. Crowe

ISBN 978-1-60692-289-7

Chapter 2

RENEWABLE ENERGY POLICY: TAX CREDIT, BUDGET, AND REGULATORY ISSUES*

Fred Sissine

ABSTRACT

High gasoline and natural gas prices have rekindled interest in the role that renewable energy may play in producing electricity, displacing fossil fuel use, and curbing demand for power transmission equipment. Also, worldwide emphasis on environmental problems of air and water pollution and global climate change, the related development of clean-energy technologies in western Europe and Japan, and technology competitiveness may remain important influences on renewable energy policymaking.

The Bush Administration's FY2007 budget request for the Department of Energy's (DOE' s) Renewable Energy Program seeks $359.2 million, which is $84.0 million, or 30.5%, more than the FY2006 appropriation. In support of the President's proposal for an Advanced Energy Initiative (AEI), the request includes major funding increases for solar energy (to support the Solar America initiative) and biomass (to support the Biorefinery Initiative). The main increases are for Solar Photovoltaics ($79.5 million) and Biomass ($59.0 million). Some significant cuts were also proposed, and the request sought to eliminate earmarks.

Appropriations actions by the House and the Senate Appropriations Committee have approved most of the requested FY2007 funding increases for AEI and greatly reduced earmark funding. Compared with House-passed

* Excerpted from CRS Report RL33588, dated July 28, 2006.

funding, the Senate Appropriations Committee recommendation seeks an increase of $66.1 million (5%). Table 3 shows other differences, most notably those for Biomass and Biorefinery, Geothermal, Hydro, and Weatherization programs.

Important regulatory issues have surfaced for wind energy. A major debate has erupted over the safety and economic and environmental aspects of a proposal by Cape Wind Associates to develop an offshore wind farm near Cape Cod, Massachusetts. The parties to the debate are waiting for the results of a Department of Interior (DOI) environmental impact statement and a Coast Guard study of navigational safety aspects. Also, concern that large wind turbines may disrupt radar systems led to federal actions to halt several wind farm developments, pending the results of a study by the Department of Defense (DOD) that was due in early May 2006. In late June 2006, the Sierra Club filed suit to compel completion of the DOD radar study. An agency of the United Kingdom has studied modifications to turbines and radar systems that may help solve the problem.

Also, high gasoline prices have stimulated a DOE proposal for aggressive development of cellulosic ethanol as an alternative to gasoline and corn-based ethanol. The focus is on using biotechnology to simplify processes and reduce costs.

MOST RECENT DEVELOPMENTS

On June 29, 2006, the Senate Committee on Appropriations reported H.R. 5427, the Energy and Water Appropriations bill for FY2007 (S.Rept. 109-274). This bill includes funding for the Department of Energy (DOE) Renewable Energy Program, which is conducted by the Office of Energy Efficiency and Renewable Energy (EERE). Compared with House-passed funding, the FY2007 Senate Appropriations Committee recommendation seeks an increase of $66.1 million (5%). Table 3 shows other differences, most notably those for Biomass and Biorefinery, Geothermal, Hydro, and Weatherization programs.

Compared with FY2006 funding, the Senate Committee recommends an increase of $240.4.8 million for R&D and deployment programs. This reflects support for the Advanced Energy Initiative, including increases for Hydrogen ($34.2 million), Biomass/Biorefineries ($59.3 million), and Solar ($65.3 million). The main cuts include the Weatherization program (-$42.6 million), Facilities (-$20.2 million), Industrial programs (-$9.3 million), and Program Management (-$9.9 million).

(The renewable energy provisions in the Energy Policy Act of 2005 [P.L. 109- 58, H.R. 6] and other bills of the 109th Congress are discussed in the

"Renewables in the 109th Congress" and "Legislation" sections below. A list of all renewable energy bills introduced in the 109th Congress is provided in CRS Report RL32860, Energy Efficiency and Renewable Energy Legislation in the 109th Congress, by Fred Sissine.)

BACKGROUND

Renewable Energy Concept

Renewable energy is derived from resources that are generally not depleted by human use, such as the sun, wind, and water movement. These primary sources of energy can be converted into heat, electricity, and mechanical energy in several ways. There are some mature technologies for conversion of renewable energy such as hydropower, biomass, and waste combustion. Other conversion technologies, such as wind turbines and photovoltaics, are already well developed, but they have not achieved the technological efficiency and market penetration that many expect they will ultimately reach. Although geothermal energy is produced from geological rather than solar sources, it is often included as a renewable energy resource (and is treated as such in this report). Commercial nuclear power is not generally considered to be a renewable energy resource. [1]

Contribution to National Energy Supply

According to the Energy Information Administration's (EIA' s) Annual Energy Outlook 2006, renewable energy resources (excluding wood used for home heating) supplied about 5.7 Q (quadrillion Btu's, or quads) of the 99.7 Q the nation used in 2004, or about 6.0% of national energy demand. More than half of renewable energy production takes the form of electricity supply. Of this, most is provided by large hydropower. However, from 1998 through 2001, a drought-driven decline in hydroelectric availability led to a major drop in national renewable energy use. Industrial use of renewables, supplied primarily by biofuels, accounts for most of the remaining contribution. [2]

Some note that renewable energy, after more than 25 years of federal support, has achieved neither a high level of market penetration nor a growing market share among other energy sources. A 1999 review of renewable energy studies by Resources for the Future (Renewable Energy: Winner, Loser, or Innocent Victim?) concludes that the lower-than-projected market penetration and flat market share were due primarily to declining fossil fuel and electricity prices during most of that period. In contrast, however, it notes that the costs for renewable energy technologies have declined by amounts equal to or exceeding those of earlier projections. [3] Nevertheless, the costs are still higher than those for conventional energy. [4] Even wind energy, which has a much lower cost than solar energy technologies, owes its recent commercial success to support from the federal renewable energy production tax credit and the renewable portfolio standard (RPS) that many states have put in place. [5]

EIA's Annual Energy Outlook 2006 projects that current policies would yield a 1.8% average annual increase in renewable energy production to 9.0 Q through 2030, resulting in a 57% total increase. This would amount to about 6.7% of the projected 134 Q total demand in 2030. [6]

Role in Long-Term Energy Supply

Our Common Future, the 1987 report of the United Nations World Commission on Environment and Development, found that "energy efficiency can only buy time for the world to develop 'low-energy paths' based on renewable sources." [7] Though many renewable energy systems are in a relatively early stage of development, they offer "a potentially huge primary energy source, sustainable in perpetuity and available in various forms to every nation on Earth." The report suggested that a research, development, and demonstration (RD&D) program of renewable energy projects is required to attain the level of primary energy now obtained from a mix of fossil, nuclear, and renewable energy resources.

The Agenda 21 adopted at the 1992 United Nations Conference on Environment and Development (UNCED, "Earth Summit") concluded that mitigating urban air pollution and the adverse impact of energy use on the atmosphere — such as acid rain and climate change — requires an emphasis on "clean and renewable energy sources." The U.N. Division on Sustainable Development oversees implementation of Agenda 21. The 2002 U.N. World Summit on Sustainable Development (Johannesburg Earth Summit) adopted a Political Declaration and a Plan of Implementation, [8] which includes "Clean Energy" as one of five key policy actions. [9] Subsequently, in 2003, the U.S. Department of State implemented a $42 million Clean Energy Initiative [10] and the European Union committed to a $700 million energy partnership. Nevertheless, EIA estimates that renewable energy's share of global energy consumption through 2030 will remain limited to less than 9% for the next two to three decades. [11]

History

The oil embargo of 1973 sparked a quadrupling of energy prices, major economic shock, and the establishment of a comprehensive federal energy program to help with the nation's immediate and long-term energy needs. During the 1970s, the federal renewable energy program grew rapidly to include basic and applied R&D and federal participation with the private sector in demonstration projects, commercialization, and information dissemination. In addition, the federal government instituted market incentives, such as business and residential tax credits, and created a utility market for nonutility-produced electric power through the Public Utility Regulatory Policies Act (P.L. 95-617).

The subsequent failure of the oil cartel and the return of low oil and gas prices in the early 1 980s slowed the federal program. Despite Congress's consistent support for a broader, more aggressive renewable energy program than any Administration, federal spending for these programs fell steadily through 1990. Until 1994, Congress led policy development and funding through legislative initiatives and close reviews of annual budget submissions. FY1995 marked a noteworthy shift, with the 103rd Congress for the first time approving less funding than the Administration had requested. The 104th Congress approved 23% less than the Clinton Administration request for FY1 996 and 8% less for FY1 997. However, funding turned upward again during the 1 05th and 1 06th Congresses, and stayed relatively flat through the 1 07th and 108th Congresses and the first session of the 109th Congress. [12]

From FY1978 through FY2005, the DOE spent about $13.4 billion (in 2005 constant dollars) for renewable energy R&D. Renewable energy R&D funding grew from less than $1 million per year in the early 1970s to more than $1.4 billion in FY1979 and FY1980, then declined steadily to $148 million in FY1990. By FY2005, it reached $380 million in 2005 constant dollars.

This spending history can be viewed within the context of DOE spending for the three major energy supply R&D programs: nuclear, fossil, and energy efficiency R&D. From FY1948 through FY1977 (in 2005 constant dollars), the federal government spent about $42.6 billion for nuclear (fission and fusion) energy R&D and about $14.1 billion for fossil energy R&D. From FY1978 through FY2005, the federal government spent $33.9 billion for nuclear (fission and fusion), $21.1 billion for fossil, $13.4 billion for renewables, and $12.4

billion for energy efficiency. Thus, total energy R&D spending from FY1948 to FY2005, in 2005 constant dollars, reached $140.0 billion, including $76.5 billion, or 55%, for nuclear; $35.2 billion, or 25%, for fossil; $13.4 billion, or 10%, for renewables; and $12.4 billion, or 9%, for energy efficiency.

In FY2006, as part of a restructuring of the appropriations committees, Congress merged appropriations accounts for the DOE Energy Efficiency Program, which previously had been under the Department of Interior and Related Agencies Appropriations Bill, with the appropriations accounts for the DOE Renewable Energy Program under the Energy and Water Development Appropriations Bill. As a result, appropriations for some subprograms of the Renewable Energy Program, such as Hydrogen and Program Management, are commingled with those for the Energy Efficiency Program and are no longer reported separately. In place of the former totals for the Renewable Energy Program, this report now shows a subtotal of appropriations for all of the renewable energy technology subprograms, namely Biomass and Biorefinery, Solar, Wind, Geothermal, and Small Hydro subprograms (see Table 3). The total for these subprograms was $242.5 million for FY2005 and $236.2 million for FY2006.

Tax Credits. The Energy Tax Act of 1978 (P.L. 95-618) created residential solar credits and residential and business credits for wind energy installations; it expired on December 31, 1985. [13] However, business investment credits were extended repeatedly through the 1980s for solar, geothermal, and ocean thermal energy technologies. Section 1916 of the Energy Policy Act of 1992 (P.L. 102-486) extended the 10% business tax credits for solar and geothermal equipment indefinitely. [14] Also, Section 1914 created a "production" tax credit (PTC) of 1.5 cents/kwh (adjusted annually for inflation) for electricity produced by wind and closed-loop biomass. [15] P.L. 106-170 expanded this credit to include poultry waste. The JOBS Act (P.L. 108-357) expanded the PTC (adding solar, geothermal, and open-loop biomass, landfill gas, trash combustion, and certain small hydro) and extended it through the end of 2005. In August 2005, the Energy Policy Act of 2005 (EPACT, P.L. 109-58) made the PTC retroactively effective to the beginning of 2005 and extended it through the end of 2007. In addition, P.L. 96-223 created an income tax credit for alcohol fuels; section 9003(a)(3) of P.L. 105-178 extended the 40-60 cents/gallon credit through December 31, 2007. Further, the Energy Tax Act created a 4.0 cents/gallon federal excise tax exemption for gasohol (gasoline blended with alcohol). The JOBS Act (P.L. 108-357) reformed this subsidy, converting the exemption to an equivalent tax credit, which now stands at 5.1 cents/gallon for a 10% ethanol blend.

Public Utility Regulatory Policies Act. The Public Utilities Regulatory Policies Act (PURPA, P.L. 95-6 17) required electric utilities to purchase power produced by qualified renewable power facilities. Under PURPA, the Federal Energy Regulatory Commission (FERC) established rules requiring that electric utilities purchase power from wind farms and other small power producers at an "avoided cost" price based on energy and capacity costs that the utility would otherwise incur by generating the power itself or purchasing it elsewhere. However, Section 1253 of P.L. 109-58 terminated the mandatory purchase and sale requirements for a new renewable power facility, provided that FERC finds that the new facility has access to wholesale power markets and transmission services.

State and Local Government Roles. State and local governments have played a key role in renewable energy development. For example, in the early 1 980s, a generous state investment tax credit for wind energy in California combined with PURPA and the federal tax credit to stimulate industry development of the first wind farms. California and New York have invested some state funds in renewable energy R&D. Recently, Texas and about 20 other states have used a regulatory tool, the renewable energy portfolio standard (RPS), to encourage renewable energy. Also, in 2001, the city of San Francisco enacted a $100 million revenue bond (Proposition B, "Vote Solar") to support solar and wind energy implementation. In 2004, the city of Honolulu approved a $7.85 million solar and energy efficiency bond. [16]

RENEWABLES IN THE 109TH CONGRESS

Action in the Second Session

DOE Budget, FY2007. On February 6, 2006, President Bush issued the Administration's budget request for FY2007. The DOE request seeks $359.2 million for renewables, which is $84.0 million, or 30.5%, more than the FY2006 appropriations (excluding inflation). The Administration's request includes funding for an Advanced Energy Initiative (AEI) as part of its American Competitiveness Initiative (ACI), which includes accelerated funding for the Solar America and Biorefinery initiatives under DOE's Renewable Energy Program. For more budget details, see "DOE Budget, FY2007" and Table 3. [17]

The House Appropriations Committee report (H.Rept. 109-474) includes several policy directives to the Office of Energy Efficiency and Renewable Energy (EERE). First, it says (pp. 72-73) that EERE could have avoided employee layoffs at the National Renewable Energy Laboratory (NREL) through better management of uncosted balances, and directs EERE to report by January 31, 2007, on steps taken to identify prior year balances and account for all out-year commitments. Second, the report directs (p. 73) EERE to report by January 31, 2007, on the progress of implementing the Inspector General's recommendations (IG audit report DOE/IG0689) to improve the management of cooperative agreements. Further, the report directs (pp. 74-75) EERE to fully fund a biomass R&D grant to Natureworks LLC, strengthen recruiting from Historically Black Colleges and Universities, and to prepare a report on solar water heaters by January 31, 2007, that covers potential energy savings, market impediments, and deployment strategy. Also, one DOE-wide directive that would directly affect EERE involves funding for the Asia Pacific Partnership (APP), which would support clean, energy-efficient technologies. The report directs (pp. 67-68) DOE to submit a reprogramming request if it intends to support APP with FY2006 funds and to submit a detailed budget justification (which would be considered by the conference committee) if it proposes to use FY2007 funds.

The Senate Appropriations Committee report (S.Rept. 109-274) on Energy and Water Development Appropriations for FY2007 includes several policy directives to EERE for the Renewable Energy Program. First, it recommends (p. 116) that DOE complete unfinished awards for biorefineries before funding new ones. It urges that DOE focus on cellulosic ethanol to reduce oil imports, and directs DOE to recommend ways to implement the cellulosic biomass production incentive in EPACT (P.L. 109-58, §942). Second, the Committee joins with the House in requiring (p. 117) a report on solar water heaters. Third, it urges (p. 117) DOE to focus on non-silicon materials and directs DOE to prepare a report by March 31, 2007, on short- and long-term silicon market conditions and the potential impact on the photovoltaic market. Fourth, it recommends (p. 117) a $9 million increase to support deployment of a solar-hydrogen pilot plant that would fulfill certain sections of EPACT. Fifth, the Committee directs (p. 117) that funding for a 1 MW solar thermal facility can only be used for deployment in New Mexico. Sixth, it requests (pp. 117-118) that EERE and the Office of Electricity (OE) provide a report by March 2007 that identifies the most promising locations for wind resources and the best opportunities for integrating these potential power generation facilities into the electric grid. Seventh, it encourages (p. 118) DOE to form an interagency group to promote renewable energy use in all aspects of federal agency operations, especially those on federal lands. In particular, this group should address the issue of wind energy project delays due to Department of Defense concerns about radar interference. Eighth, it recommends (p. 118) that $2.4 million be provided as a competitive award for development of a 2 MW permanent magnet motor wind turbine, which has the potential to eliminate the need for gearboxes. Ninth, it directs (p. 118) that funding for Hydropower include a study of advanced techniques for ocean energy, including an assessment of locations for demonstration plants, with a report by May 1, 2007. Tenth, it directs (pp. 118-119) DOE to study possible impacts of plug-in hybrids on electricity supply and distribution networks, including urban areas, and to study environmental aspects of fuel-switching. Eleventh, it directs (p. 119) DOE to provide a strategy to accelerate the development of zero energy buildings by five to seven years.

U.S. Department of Agriculture (USDA) and Other Appropriations Action, FY2007. At least three other appropriations bills include some funding for renewable energy programs. These actions are summarized in Table 1, below, and program descriptions follow. First, the Farm Security Act of 2002 (§9006) created the USDA Renewable Energy Program to provide grants and loans to agriculture producers and rural small businesses for renewable energy systems and energy efficiency improvements. Second, under the Department of State, the Clean Energy program was created to support the use of renewable energy and energy efficiency to reduce greenhouse gas emissions in developing countries.

Table 1. Other Appropriations Bills ($ millions)

Department/Agency	Bill Number	FY2006 Appns.	FY2007 Request	FY2007 House	FY2007 Senate
Agriculture	H.R. 5384	23.0	10.2	20.0	25.0
State	H.R. 5522				
Clean Energy		180.0	—	—	180.0
Exports		—	—	—	3.0
Asia-Pacific Partnership		—	26.0	0.0	26.0
Defense	H.R. 5631	—	—	6.3	—

In addition, the Clean Energy Technology Exports Initiative (CETEI) was created by language in the conference report on the FY2004 Energy and Water Development Appropriations bill.18 Further, in January 2006, the Bush Administration announced that it would seek $50 million in FY2007 for the Asia Pacific Partnership on Clean Development and Climate. Third, the Defense

Appropriations Bill would fund a wind turbine demonstration project at a military base.

Offshore Wind Farms and the Cape Wind Issue. Major commercial wind farm developments in the United States and some European nations have expanded from land-based operations to offshore coastal areas. Offshore wind farms have grown rapidly in Europe, especially in the coastal areas of Germany, the United Kingdom, and Ireland. In the United States, proposals have been made primarily for offshore areas near Texas and the mid-Atlantic and northeast states.

A major issue erupted over the Cape Wind Associates proposal for a $1 billion offshore development that would install 450 megawatts (millions of watts) of wind turbine capacity in federal waters, the Horseshoe Shoal area, near Cape Cod in Massachusetts. [19] Cape Wind made the initial project application to the U.S. Army Corps of Engineers in November 2001. The development would include 130 wind turbines, each rated at 3.6 megawatts, reaching 420 feet high (with the hub 260 feet above the surface of the water), and spread over a 24 square-mile area about six miles offshore. [20] The Corps of Engineers prepared a draft environmental impact statement that was released in November 2004. [21] In August 2005, the Energy Policy Act (EPACT, §388) shifted regulatory responsibility to the Minerals Management Service at the Department of Interior. [22]

Cape Wind Associates and other proponents say the project is a safe, clean way to develop renewable energy and create jobs. Cape Wind lists supporters that include local towns and citizen groups, renewable energy groups, labor organizations (including the Seafarers International Union), academic and scientific leaders, and college organizations. Other proponents include national environmental groups, such as the Natural Resources Defense Council (NRDC) [23] and the Union of Concerned Scientists. [24] In regard to energy development, they claim that the project would reduce air pollutants and greenhouse gases that would otherwise be generated by conventional power plants. Regarding jobs, they contend that dozens of construction jobs will be created for an 18-month period, with spin-off benefits to local businesses, and that afterwards, year-round jobs will be created to monitor, operate, and maintain the wind farm. In regard to the environment, they claim that from the shore, there will be only a minimal impact on the viewshed. Regarding the tourism industry, they argue that new opportunities will be created for up-close, “majestic views” of the wind towers, which will only be possible from tour boats. Further, they argue that charter fishing boats, sailboat tours, and scenic cruises may include the wind farm as part of the Cape Cod experience. [25]

Opponents of the Cape Wind project have collaborated to create the Alliance to Protect Nantucket Sound, which includes several local towns, business and tourism organizations, commercial fishing groups, recreational fishing and boating groups, and boating safety and navigation organizations. [26] The opponents say that the proposed development poses dangers to the area's ecosystem, maritime navigation, and the Cape Cod tourism-based economy. Regarding the viewshed, the Alliance notes that many land-based landmarks smaller than the wind turbines — including a light house, radio antenna, and water towers — are easily seen from the project site. In regard to birds, it stresses that the area is one of the most important migratory staging areas on the East Coast, providing habitat for some endangered species. Regarding fishing, it says that it is one of the richest fishing grounds on the East Coast, many local fishermen make 60% of their income on Horseshoe Shoal, and no fishing or boating group supports the project. The Alliance cites a study that found tourist-related spending would be cut by nearly $200 million per year. Also, the group reports that commercial ferry lines see serious navigation hazards, local airport commissions see safety concerns, and the wind turbines may interfere with radar systems. [27]

Under its new authority, [28] MMS became the lead federal agency regulating the development of the proposed Cape Wind project. [29] As directed by EPACT (§388), MMS is preparing an environmental impact statement (EIS). [30] As part of the EIS activity, MMS is considering potential alternatives to the Cape Wind proposal, such as modifying the size of the development, phasing it, reconfiguring it, and relocating it to alternative sites. [31]

Section 414 of the Coast Guard and Maritime Transportation Act of 2006 (P.L. 109-24 1, H.R. 889) directs the Commandant of the Coast Guard to "specify the reasonable terms and conditions the Commandant determines to be necessary to provide for navigational safety" for the proposed Cape Wind energy facility in Nantucket Sound. [32] Further, it directs MMS to incorporate these terms and conditions into any lease, easement, or right-of-way for the facility. [33]

Wind Disruption of Radar Issue. There is a concern that tall wind turbines may create false signals that would be picked up by military and civilian radar. [34] Section 358 of the Defense Authorization Act for FY2006 (P.L. 109-163, H.R. 1815) requires the Department of Defense (DOD) to submit a report to Congress on

> ... the effects of windmill farms on military readiness, including an assessment of the effects on the operations of military radar installations ... and of

> technologies that could mitigate any adverse effects on military operations identified. [35]

The law was enacted on January 6, 2006, and Section 358 specifies that the report is due 120 days after enactment, which would have occurred in early May 2006. As of the date of this CRS report, the DOD report had not been issued. [36]

On March 21, 2006, the DOD and the Department of Homeland Security (DHS) issued an Interim Policy on Proposed Windmill Farm Locations, which states that

> [t]he DOD/DHS Long Range Radar Joint Program Office Interim Policy is to contest any establishment of windmill farms within radar line of site of the National Air Defense and Homeland Security Radars. This is to remain in effect until the completion of the study and publishing of the Congressional Report. [37]

The American Wind Energy Association (AWEA) contends that many wind farm projects have received a Notice of Presumed Hazard from the Federal Aviation Administration (FAA) and have halted development. This, it says, includes projects in Wisconsin, Illinois, North Dakota, South Dakota, and Minnesota. [38] Defense and FAA officials say the "proposed hazard" letters do not prohibit wind farms, they just delay them until any risks to military operations can be assessed and resolved. [39] However, the wind industry is anxious about any delay in construction because the renewable energy production tax credit is set to expire at the end of 2007. [40] On June 2, six Senators from Illinois, Wisconsin, and North Dakota sent letters of concern to DOD about the interim policy of halting wind farms classified as within "radar line of sight" while the study is being completed. [41] On June 28, the Sierra Club filed suit against DOD for delaying the study of wind farm impact on radar mandated by the Defense Authorization Act. [42]

The British Department of Trade and Industry (DTI) has supported studies of the radar problem for wind farms in the United Kingdom. A 2003 DTI study found that there is a potential for wind turbines to significantly affect the performance of air traffic control radar. The two general categories of effects are reports of false targets (aircraft) to the radar operator and a reduction in the likelihood of detecting aircraft. This study recommended that the problem could be addressed by modifying the design and installation of both the turbines and the radar systems. [43] A 2005 DTI study modeled the use of radar absorbent materials (RAM) in wind turbine blades. [44] It found that RAM materials can be developed with minimal structural impact, [45] and recommended that a turbine outfitted with such materials be field-tested. Also, DTI formed a partnership with the British Wind Energy Association (BWEA) to fund research on an advanced digital tracker to mitigate the radar problem. Both the Ministry of Defence and the National Air Traffic Services began planning flight trials in late 2005. [46] In 2006, the two agencies decided to conduct tests with Royal Air Force (RAF) aircraft in a location in Wales near four working wind farms. [47]

Energy Policy Act of 2005 (EPACT, H.R. 6, P.L. 109-58)

Renewable Energy Portfolio Standard (RPS). For retail electricity suppliers, an RPS sets a minimum requirement (often a percentage) for electricity production from renewable energy resources or for the purchase of tradable credits that represent an equivalent amount of production. A growing number of states have enacted an RPS, currently including 19 states and the District of Columbia. [48]

The Senate Committee on Energy and Natural Resources held a hearing on RPS on March 8, 2005. Regional differences in the availability of renewable resources, particularly resource availability in the southeastern United States, was a key issue of the discussion. In the markup of a committee print (to be incorporated into H.R. 6) by the House Committee on Energy and Commerce, an amendment to add an RPS (1% in 2008, increasing by 1% annually through 2027) was rejected (17-30). Proponents noted a growing number of states with an RPS and that EIA reports show an RPS could reduce electricity bills. [49] Opponents raised concerns about the exclusion of existing hydropower facilities and resource limits for the southeastern United States.

There was no RPS provision in the House version of H.R. 6. The Senate version had a 10% RPS provision. During the conference, there was an idea put out to compromise by including nuclear and hydropower facilities. Nevertheless, RPS was dropped in conference. [50]

Renewable Energy Production Tax Credit (PTC). The House version of H.R. 6 had no PTC extension, the Senate version had a three-year extension, and the enacted law (§1301) extends the PTC for two years, through the end of calendar year 2007. [51] For claims against 2005 taxes, the credit is valued at about 1.9 cents per kilowatt-hour (kwh) for electricity produced by facilities that use wind, closed-loop biomass, poultry waste, geothermal energy, or solar energy. Also, half credit (valued at about 0.9 cents/kwh in 2005) is available for electricity produced by facilities that use open-loop biomass, incremental hydropower, municipal solid waste (landfill gas), or small irrigation water flows.

Renewable Energy Production Incentive (REPI). Parallel to the PTC, there is a renewable energy production "incentive" (REPI) for state and local governments and nonprofit electrical cooperatives. This 1.5 cent/kwh incentive was created by the Energy Policy Act of 1992 (P.L. 102-486) §1212 and is funded by appropriations to DOE. Eligible facilities currently include solar, wind, biomass (except municipal solid waste), and geothermal energy (except certain types of dry steam geothermal energy). The Energy Policy Act of 2005 (§202) expands REPI to include ocean and wave energy and extends the authorization through FY20 16.

Renewable Fuel Standard (RFS). P.L. 109-58 (H.R. 6, § 1501) defines "renewable fuel" to include ethanol, biodiesel, and natural gas produced from landfills, sewage treatment plants, and certain other sources. Ethanol is the renewable motor fuel produced in greatest quantity. In 2004, about 3.9 billion gallons of ethanol were blended with gasoline. [52] In 2005, about 75 million gallons of biodiesel were used. In the House version of H.R. 6, the RFS provision called for renewable fuels (primarily ethanol) production to reach 3.1 billion gallons a year in 2005, and then increase stepwise to 5.0 billion gallons a year by 2012. In the Senate version of H.R. 6, the RFS called for 4.0 billion gallons in 2006, rising to 8.0 billion gallons in 2012. The enacted version (§1501) set a standard starting at 4.0 billion gallons in 2006 and rising to 7.5 billion gallons by 2012. Further, an increased incentive would encourage the use of cellulosic and waste-derived ethanol to help fulfill the RFS target. [53] The previous incentive was created by crediting 1.0 gallon of cellulosic or waste-derived ethanol as equivalent to 1.5 gallons of renewable fuel. Under EPACT, it becomes equivalent to 2.5 gallons of renewable fuel.

Renewable Hydrogen. P.L. 109-58 (H.R. 6, §933) would create a program to produce hydrogen from a variety of sources, including renewable energy and renewable fuels, as part of a broader effort to develop hydrogen fuels, vehicles, and infrastructure. The provision includes a focus on distributed energy that uses renewable sources. Another section (§8 12) also calls for use of renewable hydrogen as part of a hydrogen fueling and infrastructure demonstration program.

Renewables Tax Revenue Effect. Table 2 shows the estimated 11-year revenue effect of renewable energy tax provisions in the House version of H.R. 6 (H.R. 1541), the Senate version of H.R. 6, and the enacted law.

Table 2. H.R. 6, Tax Revenue Effect ($ billions)

	House	Senate	Conference
Renewable Energy Production Tax Credit (PTC)	—	$4.577	$2.747
Clean Renewable Energy Bonds	—	$0.493	$0.41 1
Business Investment Tax Credit (Solar and Geo.)	—	$0.059	$0.024
Residential Solar Tax Credit (includes fuel cells)	$0.0 18	—	$0.03 1
Biodiesel Tax Credit	—	$0.402	$0.194
Total, Renewables Provisions	$0.018	$5.53 1	$3.407
Gross Total, All Tax Cut Provisions	$8.090	$18,421	$14.553
Renewables Share of Total	0.2%	30.0%	23.4%

Sources: Joint Committee on Taxation (JCT), Estimated Budget Effects of the Conference Agreement for Title XIII of H.R. 6, July 27, 2005 (JCX-59-05); Estimated Revenue Effects of the Chairman's Amendment in the Nature of a Substitute to H.R. 1541, Scheduled for Markup by the Committee on Ways and Means, April 13, 2005 (JCX-17-05); Estimated Revenue Effects of the Chairman's Amendment in the Nature of a Substitute to the "Energy Policy Tax Incentives Act of 2005," Scheduled for Markup by the Committee on Finance, June 16, 2005 (JCX-47-05).

Other Renewables Provisions. P.L. 109-58 (H.R. 6) covers additional areas of renewable energy policy, resources, and technology, including distributed energy, federal purchases, federal lands, Indian energy, net metering, alternative fuels (alcohol, biofuel, biodiesel), biopower/biomass, geothermal, hydropower, solar, and wind. [54]

RENEWABLE ENERGY PRODUCTION

Electricity Production from Renewables

The Public Utility Regulatory Policies Act (PURPA) was a key to the growth of electric power production from renewable energy facilities. [55] After 1994, state actions to restructure [56] the electric utility industry dampened PURPA's effect. [57] At the same time, however, the renewable energy production tax credit (PTC) grew in importance as a major incentive, helping to spur the development of independent wind energy power producers. [58] In the 109th

Congress, P.L. 109-58 (§1253) includes a conditional repeal of the mandatory renewables purchase requirement in Section 210 of PURPA. [59]

Renewables Under Electric Industry Restructuring. To encourage a continued role for renewable energy under restructuring, some states and utilities have enacted measures such as a renewable energy portfolio standard (RPS), public benefits fund (PBF), and/or "green" pricing and marketing of renewable power. The above section on "Renewable Portfolio Standard" summarizes the RPS action in H.R. 6, including a Senate proposal that was rejected in conference committee. Since the enactment of P.L. 109-58, several bills with an RPS provision have been introduced, including H.R. 4384 (§205), H.R. 5331 (§504), H.R. 5642 (§4), S. 2435 (§6b), S. 2571 (§504), and S. 2747 (§301).

Green Power. The term "green power" generally refers to electricity supplied in whole or in part from renewable energy sources. Green power marketing (retail or wholesale) is underway in California, Illinois, Massachusetts, New Jersey, New York, Pennsylvania, and Texas. Green pricing is an optional utility service that allows electricity customers who are willing to pay a premium for the environmental benefits of renewable energy to purchase green power instead of conventional power. Utility green pricing programs are available to more than one-third of the nation's consumers. [60]

Distributed Energy. Distributed energy involves the use of small, modular electricity generators sited close to the customer load that can enable utilities to defer or eliminate costly investments in transmission and distribution system upgrades, and provide customers with quality, reliable energy supplies that may have less environmental impact than traditional fossil fuel generators. Technologies for distributed electricity generation use wind, solar, bioenergy, fuel cells, gas microturbines, hydrogen, combined heat and power, and hybrid power systems. A DOE study, Structural Vulnerability of the North American Power Grid, suggests that adding more distributed power generation could help reduce grid vulnerability. [61] Another DOE study, Homeland Security: Safeguarding America's Future with Energy Efficiency and Renewable Energy Technologies, [62] provides a broad look at the potential to address vulnerabilities. [63]

Net Metering. Net metering allows customers with generating facilities to "turn their electric meters backwards" when feeding power into the grid; they receive retail prices for the excess electricity they generate. This encourages customer investment in distributed generation, which includes renewable energy equipment. About 40 states have some form of net metering in place. P.L. 109-58 (§1251) provides for net metering.

Natural Gas and Renewables

In January 2005, the Senate Energy and Natural Resources Committee held a natural gas conference. Some participants described the potential for renewable energy to augment gas supplies, reduce gas demand, and thereby help reduce natural gas prices. [64] Some of these statements referred to a 2005 DOE study entitled Easing the Natural Gas Crisis: Reducing Natural Gas Prices through Increased Deployment of Renewable Energy and Energy Efficiency. [65]

Biomass-Generated Synthetic Natural Gas (Syngas). Continuing high natural gas prices have created interest in using renewables to dampen natural gas demand. Renewable energy (mainly biomass) can be used to produce methane (the main component of natural gas), which could possibly substitute directly for natural gas. DOE projects that by 2020, biomass and energy crops could produce 15% of natural gas needs. A 2005 Harvard University study, The National Gasification Strategy, [66] cites a Princeton University study, A Cost-Benefit Assessment of Biomass Gasification Power Generation in the Pulp and Paper Industry, [67] that says that biomass-generated "black liquor" and wood waste could produce enough syngas to support 25 billion watts (gigawatts) of natural gas-fired power plant capacity by 2020.

Substituting Electricity from Renewables for Gas-Fired Generation. Also, a variety of renewables can generate electricity that indirectly displaces natural gas use for power generation. For many utilities, the peak demand (often supported with natural gas peak-load plants) occurs during hot summer afternoons. In many regions, solar and wind energy reach high levels during summer peak periods. The American Wind Energy Association (AWEA) says that at the end of 2005, wind farms were saving more than 0.5 billion cubic feet (Bcf) of natural gas per day. DOE's 2000 report Scenarios for a Clean Energy Future projects that, [68] with some federal policy changes, biomass-based power production could be greatly accelerated through 2010. [69]

Transportation Fuels Produced from Renewables

Corn Ethanol. In the United States, ethanol is produced mainly from corn [70] and is most frequently used as a 10% blend with gasoline. Historically, high cost has been a key barrier to commercial use, which has been addressed mainly by a 51-cent per gallon tax credit for fuel use. [71] Due primarily to the use of fertilizer, which has natural gas as a key ingredient, there has been a debate over the net energy benefit of using corn ethanol. The perception that ethanol use has significantly lower emissions of some air pollutants than gasoline helps maintain support for its incentives. [72]

However, compared with gasoline, ethanol can have higher emissions of volatile organic compounds (VOCs) per mile traveled. [73] Assuming that the net energy benefit of ethanol is clearly positive, then its carbon dioxide emissions are significantly lower than those for gasoline. [74]

National ethanol production was estimated at 4.0 billion gallons in 2005. [75] Due to ethanol's lower heat content, [76] this is equivalent to about 2.7 billion gallons of gasoline that year, or about 170,000 barrels of oil per day. In comparison, in 2005, U.S. transportation oil use averaged 13.82 million barrels per day (mb/d) and total U.S. oil use averaged 20.7 mb/d. [77] Thus, in 2005, ethanol production accounted for about 1.2% of transportation fuel use and about 0.8% of total oil use.

EPACT (P.L. 109-58, § 1501) established a renewable fuel standard that sets a target to increase renewable fuel production to 7.5 billion gallons per year by 2012. This target is expected to be met primarily with ethanol, and it would be equivalent to about 330,000 barrels of oil per day (b/d). Further, EIA estimates ethanol production may reach 700,000 b/d by 2030. [78] This would be equivalent to about 470,000 b/d of oil use. In comparison, EIA estimates that total U.S. oil use will reach about 25 mb/d. [79] Using these estimates suggests that ethanol would displace about 1.9% of total oil use in 2030.

Cellulosic Ethanol. Corn and other starches and sugars are only a small fraction of biomass that can be used to make ethanol. Advanced bioethanol technology could allow fuel ethanol to be made from cellulosic (plant fiber) biomass, such as agricultural and forestry wastes, industrial waste, municipal solid waste (organic material), trees, and grasses. [80] Cellulosic ethanol can be made through gassification, thermochemical, and microbial processes. In the microbial process, cellulosic ethanol is made by breaking apart the cellulose (fiber) in plant cell walls into component sugar molecules, which microbes then convert to ethanol. [81]

In particular, ethanol can be produced from dedicated fuel crops, such as fast-growing trees and switchgrass. The latter is a native grass that grows well on cropland and marginal lands, needing little water and no fertilizer. Because fertilizer is not needed, the net energy production is much higher than that for corn ethanol. Due to its capacity for production on marginal lands, the potential growing area for switchgrass is much larger than that for corn. A provision in the Farm Security Act of 2002 (P.L. 107-17 1) encourages use of the Conservation Reserve Program lands for development of such biomass resources. [82]

Although cellulosic feedstocks may be cheaper and more plentiful than corn, they require more extensive processing that makes the conversion to ethanol more costly. Both DOE and USDA are conducting research to improve this technology and reduce costs. In particular, DOE is studying the use of enzymes and heat as catalysts to reduce the cost of the conversion process. Both the United States and Canada have pilot facilities for producing cellulosic ethanol, and Canada has the only commercial-scale plant in operation. [83]

The Administration's Biorefinery Initiative, part of its Advanced Energy Initiative (AEI), aims to increase funding for cellulosic ethanol development with the goal of accelerating its commercial use. [84] In response to the AEI, DOE set a goal of using biofuels to displace 30% of national transportation fuel use (at the 2004 level) by 2030. In July 2006, DOE announced a partnership between EERE and the Office of Science to pursue use of biotechnology to produce cellulosic ethanol. A report on the research strategy of the partnership says that the United States could produce up to 60 billion gallons of biofuels per year, which would be equivalent to about 2.6 million barrels of oil per day. [85] The research plan is focused on achieving breakthroughs in biotechnology to increase the quantity of biomass (e.g., corn plant waste and switchgrass) per acre and breeding the plants to have less lignin and more cellulose. The plan also aims to achieve biorefinery breakthroughs that would reduce the number of steps in the conversion process and shift the process focus from chemical steps to biological steps in order to reduce waste byproducts and cut costs. [86]

Biodiesel. Biodiesel has been produced and used in stationary applications (heat and power generation) for nearly a century, but its use as a transportation fuel is recent. [87] DOE notes that this alternative to petroleum-based diesel fuel is a domestically produced, renewable fuel that can be manufactured from vegetable oils, animal fats, or recycled restaurant greases. [88] Further, DOE says biodiesel is safe and biodegradable and reduces air pollutants such as particulates, carbon monoxide, hydrocarbons, and air toxics; but it also notes that biodiesel tends to increase emissions of nitrous oxides. [89] Blends of 20% biodiesel (B20) with 80% petroleum diesel can generally be used in unmodified diesel engines. Biodiesel can also be used in its pure form (B 100), but it may require certain engine modifications to avoid maintenance and performance problems and may not be suitable for wintertime use. [90] Other challenges include resource availability and the establishment of a retail distribution network. [91]

Biodiesel appears to have a net energy production much higher than that for corn ethanol and slightly higher than that for petroleum diesel. [92] U.S. biodiesel production has increased from less than 1 million gallons in 1999 to an estimated 75 million gallons in 2005. [93] The latter figure is equivalent to about 5,000 barrels of oil per day and is less than 1% of total diesel fuel use for transportation. [94] EIA projects that biodiesel production could reach the equivalent of 20,000 barrels of oil per day by 2030. [95]

The American Jobs Creation Act of 2004 (P.L. 108-357, §301) created an excise tax credit for biodiesel fuel blending. The credit is set at $1.00 per gallon of biodiesel used. [96] Because the credit is fixed on a volume (per gallon) basis, a 20% biodiesel blend (B20) would be eligible for a 20 cent per gallon credit, and a 100% biodiesel fuel (B100) would be eligible for a $1.00 per gallon credit. [97] EPACT (§1344) extended this credit through the end of 2008. Further, EPACT (§1501) created a renewable fuel standard (RFS), [98] which allows for biodiesel production to contribute to achievement of the standard. [99]

Climate Change and Renewables

Because in most cases renewable energy appears to release less carbon dioxide (CO2) than fossil fuels, [100] renewables are seen as a key long-term resource that could substitute for fossil energy sources used to produce vehicle fuels and electricity. The percentage of renewable energy substitution depends on technology cost, market penetration, and the use of energy efficiency measures to control energy prices and demand. DOE's November 2005 report U.S. Climate Change Technology Programs — Technology Options for the Near and Long Term compiles information from multiple federal agencies on more than 80 technologies. [101] For these end-use and supply technologies, the report describes President Bush's initiatives and R&D goals for advancing technology development, but it does not estimate emissions saving potentials, as some previous DOE reports on the topic had presented.

Climate Action Report — 2002 describes federal renewable energy programs aimed at reducing greenhouse gas emissions. [102] In Climate Change 2001: Mitigation, the Intergovernmental Panel on Climate Change looks at the role that renewables could play in curbing global CO2 emissions. [103]

Since 1988, the federal government has accelerated programs that study the science of global climate change and has initiated programs aimed at mitigating fossil fuel-generated CO2 and other human-generated emissions. The federal government funds programs for renewable energy as a mitigation measure at DOE, USDA, the Environmental Protection Agency (EPA), the Agency for International Development (AID), and the World Bank. The latter two agencies have received funding for renewable energy-related climate actions through Foreign Operations appropriations bills.

Because CO2 contributes the largest greenhouse gas emission impact, the reduction of CO2 has been the focus of studies of the potential for reducing emissions through renewable energy and other means. Except for biofuels and biopower, wherever renewable energy equipment displaces fossil fuel use, it will also reduce carbon dioxide (CO2) emissions, as well as pollutants that contribute to water pollution, acid rain, and urban smog. In general, the combustion of biomass for fuel and power production releases CO2 at an intensity that may rival or exceed that for natural gas. However, the growth of biomass material, which absorbs CO2, offsets this release. Hence, net emissions occur only when combustion is based on deforestation. In a "closed loop" system, biomass combustion is based on rotating energy crops, there is no net release, and its displacement of any fossil fuel, including natural gas, reduces CO2 emissions.

LEGISLATION

In the 109th Congress, more than 260 bills with provisions for renewable energy or energy efficiency have been introduced. A general description of the renewable energy provisions in these bills, including those enacted into law, is available in CRS Report RL3 2860, Energy Efficiency and Renewable Energy Legislation in the 109th Congress, by Fred Sissine. The report also groups the bills by policy and issue areas, provides a table that identifies recent action on the bills, and discusses recent action on a month-by-month basis.

Appropriations Bills in the Second Session

H.R 5384 (Bennett)

Agriculture, Rural Development, Food and Drug, Administration, and Related Agencies Appropriations Bill, 2007. Under Title III, Rural Development Programs, the bill would provide funding for the USDA Renewable Energy Program. The House approved $20 million and the Senate Appropriations Committee recommends $25 million. The Senate report language (p. 113) includes earmark recommendations. House Committee on Appropriations reported (H.Rept. 109-463) May 12, 2006, and reported Part II on May 16. Passed House, amended, May 23. Senate Committee on Appropriations reported (S.Rept. 109-266) June 22.

H.R 5427 (Hobson)

Energy and Water Development Appropriations Act, 2007. Provides funding for DOE Renewable Energy Program. The details of House and Senate action are shown in Table 3. House Committee on Appropriations reported (H.Rept. 109-474) May 19, 2006, with amendments. Passed House, amended, May 24. Senate Committee on Appropriations reported (S.Rept. 109-274) June 29.

H.R 5522 (McConnell)

Department of State, Foreign Operations, and Related Programs Appropriations Bill, 2007. In the Senate version of the bill, under Title III, Bilateral Economic Assistance, the program for Development Assistance would include three types of support for renewable energy. First, the program for Energy, Biodiversity, and the Environment would, according to the Committee report (p. 65), provide $180 million "to support policies and programs in developing countries that promote energy efficiency, renewable energy, and cleaner energy technologies...." Also, $3 million would be provided for the U.S. Agency for International Development (USAID) partnership with DOE for the hydropower Clean Energy Technology Exports Initiative (CETEI). Second, under the USAID Development Assistance program, about $160 million would be provided (p. 5 9-60) for Energy, Biodiversity, and Other Environment programs for Africa ($73 million), East Asia/Pacific ($28 million), Near East ($2 million), South Asia ($18 million), and Western Hemisphere ($52 million). Third, under Other Bilateral Economic Assistance, the bill would provide $26 million for the Asia Pacific Partnership, and further specifies (p. 73) that the "...Partnership activities will be coordinated with existing efforts to promote clean energy export and market development initiatives." House Committee on Appropriations reported (H.Rept. 109-486) June 5, 2006, with amendments. Passed House, amended, June 9. Senate Committee on Appropriations reported (S.Rept. 109-277) July 10.

H.R 5631 (Jerry Lewis)

Department of Defense Appropriations Bill, 2007. The House version of this bill would provide $6.3 million to the Office of the Secretary of Defense (OSD) for a wind energy demonstration project. House Committee on Appropriations reported (H.Rept. 109-504) June 16, 2006. Passed House, amended, June 20. Senate Committee on Appropriations reported (S.Rept. 109-292)) July 25.

Table 3. DOE Renewable Energy Budget for FY2005-FY2007 (selected programs, $ millions)

OFFICE OF ENERGY EFFICIENCY AND RENEWABLE ENERGY (EERE)	FY2005 Appn.	FY2006 Appn.	FY2007 Request	FY2007 House	FY2007 Senate Cmte.	Senate-House Percent Change
HYDROGEN TECHNOLOGY	166.8	155.6	195.8	195.8	189.9	-3.0%
Fuel Cell Technology	81.9	67.8	96.6	96.6	85.4	-11.6%
BIOMASS and BIOREFINERY	87.5	90.7	149.7	149.7	213.0	42.3%
Biochemical Platform (Cellulose)	11.1	10.4	32.8	32.8	—	—
SOLAR ENERGY	84.3	83.1	148.4	148.4	148.4	0.0%
Photovoltaics	65.8	60.0	139.5	134.5	125.5	-6.7%
Concentrating Solar	5.9	7.4	8.9	8.9	17.9	101.1%
Solar Heating and Lighting	2.4	1.5	0.0	5.0	5.0	0.0%
WIND	40.6	38.9	43.8	43.8	39.4	-10.0%
GEOTHERMAL	25.3	23.1	0.0	5.0	22.5	350.0%
SMALL HYDRO	4.9	0.5	0.0	0.0	4.0	400.0%
VEHICLE TECHNOLOGIES	161.3	182.1	166.0	172.5	180.0	4.4%
BUILDING TECHNOLOGIES	65.2	69.3	77.3	80.0	95.3	19.1%
INDUSTRIAL TECHNOLOGIES	73.4	56.9	45.6	51.6	47.6	-7.8%
DISTRIBUTED ENERGY*	59.1	0.0	—	—	—	—
FEDERAL ENERGY MGMT	19.9	19.2	16.9	18.9	16.9	-10.6%
FACILITIES and INFRASTRUC. (National Renewable Energy Lab)	11.4	26.1	5.9	15.9	5.9	-63.0%
WEATHERIZATION	325.5	316.9	225.6	335.4	266.4	-20.6%
International Renewables	6.4	3.9	2.5	4.5	2.5	-44.7%
Tribal Energy	5.5	4.0	4.0	4.0	5.0	25.3%
Renew. En. Production Incentive	5.0	5.0	4.9	4.9	4.9	0.0%
PROGRAM MANAGEMENT	115.1	111.9	102.0	102.0	101.9	0.0%
RENEWABLES, SUBTOTAL[a]	270.3	275.2	359.2	376.2	445.6	18.4%
Renewables R&D Subtotal	242.5	236.2	341.9	346.9	427.3	23.2%
RENEWABLES, EARMARKS	52.0	82.6	0.0	54.9	54.3	-1.2%
Prior Year Balances (EERE)	-5.3	—	—	—	—	—

Table 3. DOE Renewable Energy Budget for FY2005-FY2007 (selected programs, $ millions) (Continued)

EERE, TOTAL	1,234.3	1,173.8	1,176.4	1,319.4	1385.5	5.0%
Office of Electricity Delivery and Energy Reliability (OE)[b]	117.9	161.9	124.9	144.0	135.0	-6.3%

Source: DOE FY2007 Budget Request, v. 3; February 2006; H.Rept. 109-474; S. Rept 109- 274.

a R&D includes Biomass, Solar, Wind, Geothermal, Hydro. Other includes Facilities, International, Tribal, REPI.

b Funding for Distributed Energy was moved to OE.

Energy Policy Act of 2005 (EPACT, P.L. 109-58)

Title II has several renewable energy provisions: Section 202 reauthorizes the renewable energy production incentive (REPI), Title II (Subtitle C) authorizes increased hydropower at existing dams, Section 203 sets a goal for renewables use in federal facilities and fleets, and Section 206 establishes a residential renewable energy rebate program. Section 812 creates a program for using solar energy to produce hydrogen. Title IX provides funding reauthorizations for renewable energy R&D programs. Section 1253 would, under certain conditions, terminate PURPA cogeneration and small (renewable) power requirements. Title XIII has several tax incentives for renewables: Section 1301 extends the renewable energy production tax credit (PTC) for two years, Section 1303 creates $800 million in renewable energy bonds, Section 1335 creates a 30% residential solar investment credit for two years, Section 1337 increases the business solar investment credit from 10% to 30% for two years, and Sections 1345, 1346, 1347, and 1348 create or extend credits for ethanol and biodiesel fuels. Title XV (Subtitle A) has several renewable fuels provisions covering ethanol, biofuels, cellulosic biodiesel, and municipal waste. In particular, Section 1501 sets a renewable fuels standard of 7.5 billion gallons per year by 2012 for increased use of ethanol and biodiesel. Title XVII empowers DOE to provide loan guarantees for certain innovative renewable energy technology projects. Section 1826 requires a study of passive solar energy. Conference reported (H.Rept. 109- 190) July 27, 2005. Signed into law August 8.

Other Public Laws of the 109th Congress

P.L. 109-148 (H.R. 2863)

Department of Defense Appropriations Bill, 2006. Title II, regarding Operation and Maintenance, includes $4.25 million for a wind power demonstration project on an Air Force base. Conference Committee reported (H.Rept. 109-359, p. 5) December 18, 2005. Signed into law December 30, 2005.

P.L. 109-171 (S. 1932)

Deficit Reduction Act of 2005. Section 1301 amends section 9006(f) of the Farm Security Act of 2002 to set a limit of $3 million in FY2007 funding for the USDA Commodity Credit Corporation to carry out renewable energy and energy efficiency projects. Section 1402 terminates FY2007 funding authorization for the USDA Value-Added Producer Program (created by section 6401 of the Farm Security Act of 2002) to provide grants to renewable energy and energy efficiency projects. Conference reported (H.Rept. 109-362) December 19, 2005. Signed into law February 8, 2006.

Note: Three other public laws make appropriations for renewable energy programs. P.L. 109-97 (H.R. 2744), the Agriculture Appropriations Bill for FY2006, includes $23 million for USDA's renewable energy grant and loan program; P.L. 109-102 (H.R. 3057, Section 585 [a]), the Department of State's Appropriations Bill for FY2006, provides $100 million for clean (renewable) energy and energy efficiency programs that seek to reduce greenhouse gas emissions in developing countries; and P.L. 109-103 (H.R. 2419) makes appropriations for the DOE renewable energy programs. [104]

P.L. 109-59 (H.R. 3)

Transportation Equity Act: A Legacy for Users (SAFETEA-LU). Section 1113 has a volumetric excise tax credit for alternative fuels. Section 1121 on High- Occupancy Vehicle (HOV) Lanes includes provisions for alternative-fueled vehicles. Section 1208 on High-Occupancy Vehicle (HOV) Lanes includes provisions for alternative-fueled vehicles and energy-efficient vehicles. Section 3010 on Clean Fuels Formula Grant Program includes provisions for biodiesel, alcohol fuels, and fuel cells. Section 3044 supports clean fuels, and Section 6015 supports clean fuel school buses. Conference reported (H.Rept. 109-203) July 28. Signed into law August 10.

REFERENCES

[1] For further definitions of renewable energy, see the National Renewable Energy Laboratory's website information on "Clean Energy 101" at [http://www.nrel.gov/learning/].

[2] EIA has prepared a brochure entitled Renewable Energy Sources: A Consumer's Guide that shows information about the energy contribution of several renewable energy sources. It is available at [http://www.eia.doe.gov/neic/brochure/renew05/renewable.html].

[3] The report by Resources For the Future is available at [http://www.rff.org/Documents/ RFF-Resources-1 35-renewenergy.pdf].

[4] See the historical trend Cost Curves at National Renewable Energy Laboratory, Energy Analysis, at [http://www.nrel.gov/analysis/newsletters/2006_january.html]

[5] Information about the 20 states that have established an RPS is available online from the Database of State Incentives for Renewable Energy at [http://www.dsireusa.org/].

[6] Detailed breakdowns of renewable energy use appear in EIA's Renewable Energy Trends 2004, at [http://www.eia.doe.gov/cneaf/solar.renewables/page/trends/ rentrends04.html].

[7] The United Nations resolution (A/RES/42/187) that accompanied release of the report is available at [http://www.un.org/documents/ga/res/42/ares42-187.htm].

[8] The declaration is available at [http://www.johannesburgsummit.org/].

[9] More information about the United Nations' role in Energy for Sustainable Development is available at [http://www.un.org/esa/sustdev/sdissues/energy/enr.htm].

[10] The Department of State document is available at [http://www.state.gov/g/oes/sus/wssd/].

[11] EIA, International Energy Outlook 2006, Table A8 (World Consumption of Hydroelectricty and Other Renewable Energy, 1990-2030) and Table A1(World Energy Consumption, 1990-2030), at [http://www.eia.doe.gov/oiaf/ieo/ieorefcase.html].

[12] A detailed description of DOE programs appears in DOE's FY2007 Congressional Budget Request, DOE/ME-0053, v. 3, February 2006, at [http://www.mbe.doe.gov/budget/07budget/
Content/Volumes/vol_3_ES.pdf].

[13] DOE's Energy Information Administration has a historical summary of tax provisions affecting renewable energy in an article entitled Legislation

Affecting the Renewable Energy Marketplace, available at [http://www.eia.doe.gov/cneaf/solar.renewables/page/legislation/ impact.html].

[14] The credit for ocean thermal energy expired at the end of 1988.

[15] By definition, closed-loop biomass is derived from energy crops or trees grown only for use as a fuel. It does not include, for example, wood wastes from a lumber mill.

[16] For more information on federal, state, and local policies (incentives, grants, standards) for renewable energy, see Database of Incentives for Renewable Energy, at [http://www.dsireusa.org/].

[17] CRS Report RL33294, DOE Budget Earmarks: A Selective Look at Energy Efficiency and Renewable Energy R&D Programs, by Fred Sissine, provides more details about earmarks for DOE's renewable energy programs. Also, the DOE budget document is online at [http://www.cfo.doe.gov/budget/index.htm].

[18] The conference report (H.R. 5427, H.Rept. 108-357, p. 142) created an interagency group led by DOE, which was directed to provide a report to Congress by January 15, 2004, on the status of the implementation of the CETEI strategic plan and specific actions that each of the participating agencies had taken in FY2003 and planned to take in FY2004 to engage non-governmental, private sector, and other international partners. In addition, the conferees directed DOE to make $400,000 available to establish an interagency CETEI center in the Office of International Energy Market Development.

[19] A summary of the Cape Wind Project plan materials is available at [http://www.mms.gov/ offshore/PDFs/CapeWindProjectPlanFiling.pdf].

[20] Minerals Management Services (MMS). A map of the proposed location is available on the MMS website at [http://www.mms.gov/offshore/RenewableEnergy/CapeWind AlternativeSites3b.pdf]. Also, the Alliance to Save Nantucket Sound has a map posted online at [http://www.saveoursound.org/Cape/Overview].

[21] U.S. Army Corps of Engineers, "Corps to Release Draft EIS on Proposed Wind Energy Project in Nantucket Sound," news release, Nov. 8, 2004, p. 4. The Corps says it examined a broad array of "public interest factors" and considered four alternative locations; see [http://www.nae.usace.army.mil/news/2004-105.htm].

[22] EPACT transferred federal regulatory authority over the project to the U.S. Department of Interior's Minerals Management Services (MMS). MMS already had jurisdiction for oil and gas development in the offshore region of the Outer Continental Shelf, and EPACT expanded this authority to

include renewable energy development, covering wind and wave energy; see [http://www.mms.gov/offshore/RenewableEnergy/RenewableEnergy Main.htm].

[23] In a press release, NRDC stated qualified support for the project: "... it is crucial for us to set the bar [for environmental standards] in the right place. The air quality, public health, and global warming benefits of the project are significant and beyond rational dispute. It is also critical to ensure that the project will protect marine wildlife and comply with all applicable laws ... We also recommend additional steps for the project to take...." NRDC, NRDC Submits Comments on Cape Wind Project as Public Comment Period Closes, Feb. 24, 2005, at [http://www.nrdc.org/media/pressreleases/ 050225.asp].

[24] Union of Concerned Scientists (UCS), Comments on the Cape Wind Draft Environmental Impact Statement. UCS says it was encouraged by the Army Corps Draft EIS findings of the project's "significant positive" benefits and that the majority of potential negative impacts would be "minor, temporary, localized, mitigated ... or will not occur." See [http://www.ucsusa.org/ clean_energy/renewable_energy_basics/cape-wind-comments-on -draft-environmental-impact-statement.html?print=t].

[25] Cape Wind, Boosting Cape Cod's Economy, at [http://www.capewind.org/ article2.htm].

[26] A detailed list of opponents is provided by the Alliance to Protect Nantucket Sound at [http://www.saveoursound.org/Cape/Opposition].

[27] Alliance to Save Nantucket Sound. Save Our Sound. Impacts and Effects.

[28] Prior to EPACT, the U.S. Army Corps of Engineers had authority over th Cape Wind lease proposal, and the Corps had completed more than four years of reviews on the proposals, which will be considered by MMS.

[29] MMS has authority for issuing leases, easements, or rights-of-way for alternative energy projects on the Outer Continental Shelf (OCS). This authority is described in greater detail on the MMS website, available at [http://www.mms.gov/offshore/RenewableEnergy/OCSAlternativeEnergyA ndRelatedUses.pdf].

[30] The May 30, 2006, Federal Register notice (p. 30693) of MMS's intent to publish an EIS is available at [http://www.mms.gov/federalregister/ PDFs/NOI5_30_06.pdf]. MMS expects to publish a draft EIS in the winter of 2006, which will be followed by public hearings, with a final EIS expected in the fall of 2007. More information about the EIS is available at [http://www.mms.gov/ooc/press/2006/press07 13 .htm].

[31] MMS, Outer Continental Shelf, Headquarters, Cape Wind Offshore Wind Development 2007, Federal Register, v. 71, May 30, 2006, p. 30694. The MMS website home page is available at [http://www.mms.gov/offshore/RenewableEnergy/CapeWind.htm].

[32] The Coast Guard's study is being conducted by its Office of Navigational Standards.

[33] This language was inserted into the H.R. 889 by S.Con.Res. 103. It is the enacted compromise, reflecting a change from a proposed amendment to the Senate bill that would have given the Governor of Massachusetts the power to veto the project.

[34] One study suggests that this problem may occur with some, but not most, wind farms. For a description and illustration of the technical aspects of the wind-radar issue, see Idaho National Laboratory, Wind Radar Interference, by Gary Seifert, June 2006, at [http://www.eere.energy.gov/windandhydro/windpoweringamerica/pdfs/workshops/2006 _summit/seifert.pdf].

[35] P.L. 109-163, Section 358, p. 74.

[36] The DOD staff coordinator for the radar study is Mr. Andrew Kavaliunas. His phone number is 703-614-6375. The Coast Guard's study underway pursuant to P.L. 109-24 1 will also look at some radar-related issues.

[37] AWEA, Wind Turbines and Radar: An Informational Resource, Apr. 24, 2006, 3 p., at [http://www.renewwisconsin.org/windfarm/060424%20Wind%20Turbines%20and%20 Radar.pdf]. [Emphasis added.]

[38] AWEA, Wind Turbines and Radar: An Informational Resource, 3 p., at [http://www.awea.org/pubs/factsheets/060602_Wind_Turbines_and%20_Radar_Fact_Sheet. pdf]

[39] Washington Post, "Wind Power Projects Halted," June 10, 2006, p. A2.

[40] Washington Post, "A New Blip on Wind Power's Radar Screen," June 20, 2006, p. D1.

[41] Honorable Richard Durbin, Durbin and Other Midwestern Senators Object to Shut Down of Wind Farms, press release, June 2, 2006, at [http://durbin.senate.gov/record.cfm?id= 256426&&].

[42] Sierra Club, Sierra Club Calls on Rumsfeld, DOD to Stop Blocking Wind Farms, June 28, 2006, at [http://www.sierraclub.org/pressroom/releases/pr2006-06-28.asp]. The suit was filed in the U.S. District Court for the Northern District of California. It claims that DOD has violated the Administrative Procedure Act (5 U.S.C. §706[1]) and will seek to "compel agency action unlawfully withheld or unreasonably delayed."

[43] United Kindom, DTI, Feasibility of Mitigating the Effects of Windfarms on Primary Radar, 2003, Section 8, Conclusions and Recommendations, at [http://www.dti.gov.uk/ files/file17878.pdf]

[44] DTI, Technology Programme: New and Renewable Energy, Design and Manufacture of Radar Absorbing Wind Turbine Blades — Final Report, February 2005, Executive Summary at [http://www.dti.gov.uk/files/file 160 10.pdf?pubpdfdload=05%2F1409].

[45] Specifically, the study modeled a single layer of fiberglass cloth with an electromagnetically modified version of the same blade material.

[46] BWEA Press Release, Government and Industry Join Forces to Find Clear Path for Wind Farms: Trials Underway to Reduce Aviation Objections to Wind Farms, July 2005, at [http://www.bwea.com/media/news/ adt_trials.html]

[47] DTI, Government News Network, The Trials of Climate Change — RAF Fly Past, June 7, 2006, 3 p., at [http://www.gnn.gov.uk/environment/detail. asp?ReleaseID=206152&NewsAreaID=2&NavigatedFromDepartment=Fals e].

[48] The state RPS provisions are listed in the Database of State Incentives for Renewable Energy, at [http://www.dsireusa.org/]. Search by "incentive type," then "portfolio standards."

[49] See EIA's 2002 report Impacts of a 10-percent Renewable Portfolio Standard, at [http://tonto.eia.doe.gov/FTPROOT/service/sroiaf(2002)03 .pdf], and its 2003 Analysis of a 10-percent Renewable Portfolio Standard, at [http://tonto.eia.doe.gov/FTPROOT/ service/sroiaf(2003)0 1 .pdf].

[50] For more background information on RPS, two memoranda are available from the author. For current status of RPS policies in the states, see Database of State Incentives for Renewable Energy, at [http://www.dsireusa.org].

[51] A detailed description of the PTC appears in the report Description and Analysis of Certain Federal Tax Provisions Expiring in 2005 and 2006, by the Joint Tax Committee, at [http://www.house.gov/j ct/x-12-05.pdf].

[52] Renewable Fuels Association, Industry Statistics, Historic U.S. Fuel Ethanol Production 1980-2005, at [http://www.ethanolrfa.org/industry/ statistics/].

[53] Section 1501 requires the use of 250 million gallons per year of ethanol distilled from cellulosic materials, starting in 2012. EIA, Annual Energy Outlook 2006, "U.S. Demand for Ethanol Fuel Varies With World Oil Price Projections," at [http://www.eia.doe.gov/oiaf/aeo/ gas.html].

[54] Additional bills with renewable energy provisions are identified in CRS Report RL3 2860, Energy Efficiency and Renewable Energy Legislation in the 109th Congress, by Fred Sissine.

[55] PURPA required electric utilities to interconnect with renewable energy power facilities and to pay a price (avoided cost) that was determined by each state.

[56] For a discussion of broader electricity restructuring issues, see CRS Report RL32728, Electric Utility Regulatory Reform: Issues for the 1 09th Congress, by Amy Abel.

[57] Electricity restructuring aimed to create competition in electricity markets, primarily by deregulating electricity generation and opening access to utility transmission lines (retail wheeling) for all electricity producers.

[58] Further, near the turn of the century, states began to establish a renewable energy portfolio standard (RPS) that required retail electricity suppliers to produce a fixed minimum amount of electricity from renewables. The RPS, working together with the PTC, has also become a significant incentive for renewable energy power production.

[59] The conditions are that the cogeneration or small power production facility has nondiscriminatory access to — A(i) independently administered, auction-based day ahead and real time wholesale markets for the sale of electric energy; and (ii) wholesale markets for long-term sales of capacity and electric energy; or (B)(i) transmission and interconnection services that are provided by a Commission-approved regional transmission entity and administered pursuant to an open access transmission tariff that affords nondiscriminatory treatment to all customers; and (ii) competitive wholesale markets that provide a meaningful opportunity to sell capacity, including long-term and short-term sales, and electric energy, including long-term, short-term and real-time sales, to buyers other than the utility to which the qualifying facility is interconnected. In determining whether a meaningful opportunity to sell exists, the Commission shall consider, among other factors, evidence of transactions within the relevant market; or (C) wholesale markets for the sale of capacity and electric energy that are, at a minimum, of comparable competitive quality as markets described in subparagraphs (A) and (B).

[60] For more on green power, see [http://www.eere.energy.gov/greenpower/].

[61] DOE, National Transmission Grid Study, May 2002, p. 82 and 94, at [http://www.pi.energy.gov/pdf/library/TransmissionGrid.pdf].

[62] DOE, Homeland Security: Safeguarding America's Future with Energy Efficiency and Renewable Energy Technologies, Tenth Annual Report of

the State Energy Advisory Board, August 2002, 26 p., at [http://steab.org/docs/STEAB_Report_2002.pdf].

[63] In the FY2006 appropriations process, Congress directed DOE to move the Distributed Energy program from EERE to OE. More information about DOE's Distributed Energy Program is available at [http://www.eere.energy.gov/de/].

[64] See [http://energy.senate.gov/conference/conference.cfm].

[65] This report is available at [http://eetd.lbl.gov/ea/ems/reports/56756.pdf].

[66] The Harvard study is available at [http://bcsia.ksg.harvard.edu/publication.cfm?program= CORE&ctype=paper&item_id=473].

[67] The Princeton study is available at [http://www.agenda2020.org/PDF/BLGCC%20FINAL%20REPORT%208%20OCT%202003 .pdf].

[68] The projection is in Table 7.11; the report is at [http://www.ornl.gov/sci/eere/cef/].

[69] Also, see the 2005 DOE study noted above and the American Council for an Energy Efficient Economy's report Impacts of Energy Efficiency and Renewable Energy on Natural Gas Markets, available at [http://www.aceee.org/energy/efnatgas-study.htm].

[70] In other countries, such as Brazil, ethanol is often made from cane sugar.

[71] David Morris of the Institute for Local Self-Reliance says that, with recent high gasoline prices, ethanol is now price competitive. However, given the uncertainty about the stability of high prices, he proposes "that half the current 51 cent per gallon incentive be converted into an indexed incentive." See his article, The New Ethanol Future Demands a New Public Policy, June 21, 2006, at [http://www.ilsr.org/columns/2006/062 106.html].

[72] For more information about ethanol developments and issues, see CRS Report RL33564, Alternative Fuels and Advanced Technology Vehicles: Issues in Congress, and CRS Report RL33290, Fuel Ethanol: Background and Public Policy Issues, both by Brent Yacobucci.

[73] DOE. Clean Cities, Low-Level Ethanol Fuel Blends, Fact Sheet, April 2005, p. 1, at [http://www.eere.energy.gov/cleancities/blends/pdfs/37 1 35.pdf].

[74] Net greenhouse gas emissions tend to follow the net energy balance. See Farrele et al., "Ethanol Can Contribute to Energy and Environmental Goals," Science, v. 311, Jan. 27, 2006, pp. 506-508.

[75] Renewable Fuels Association, Annual Industry Outlook 2006, February 2006, p. 3, at [http://www.ethanolrfa.org/industry/outlook/]. Also, DOE reports that in 2004, 3.4 billion gallons were produced. DOE, EERE, Biomass Program: Ethanol, at [http://www 1 .eere.energy.gov/biomass/ethanol.html].

[76] DOE, EIA, Ethanol. EIA reports that the heat content of ethanol is about 3.5 million Btu per barrel (42 gallons); see [http://www.eia.doe.gov/oiaf/ethanol3.html]. Also, EIA's Monthly Energy Review, at [http://www.eia.doe.gov/emeu/mer/append_a.html], reports that the heat content of motor gasoline is 5.25 million Btu per barrel. Thus, on a per volume basis, ethanol has about 67% of the heat content of gasoline.

[77] Also, EIA estimates that total U.S. petroleum consumption averaged 20.6 million barrels per day (mb/d) in 2005. EIA, Basic Petroleum Statistics, updated July 2006, at [http://www.eia.doe.gov/neic/quickfacts/quickoil.html].

[78] EIA, Annual Energy Outlook 2006, Issues in Focus, p. 58, at [http://www.eia.doe.gov/ oiaf/aeo/index.html].

[79] EIA, Annual Energy Outlook 2006, Reference Case, Table A1, p. 58, at [http://www.eia.doe.gov/oiaf/aeo/index.html].

[80] The advanced technology has been demonstrated in the laboratory but has yet to be incorporated into a successful commercial operation. DOE, EERE, Biomass Program: Ethanol, at [http://www1.eere.energy.gov/biomass/ethanol.html].

[81] DOE, Fact Sheet: A Scientific Roadmap for Making Cellulosic Ethanol A Practical Alternative to Gasoline, July 7, 2006, 3 p., at [http://www.doegenomestolife.org/biofuels/ B2Bfactsheet.pdf].

[82] For more information about using cellulosic biomass for ethanol production, see CRS Report RL32712, Agriculture-Based Renewable Energy Production, by Randy Schnepf.

[83] Agriculture-Based Renewable Energy Production, pp. 12-14. However, in June 2006, the New York State Energy Research and Development Authority (NYSERDA) issued a request for proposals to build a pilot plant that included a requirement that the company show an ability to use the project experience to develop a commercial facility; see [http://www.ny.gov/governor/press/06/06 15063 .html]. Also, in July 2006, Xethanol Corporation announced plans to build a full-scale cellulosic ethanol plant in Georgia by mid- 2007. The plant will use forest products to produce 50 million gallons per year. DOE, EERE News, July 26, 2006.

[84] The White House, Fact Sheet: President Bush's Four-Part Plan to Confront High Gasoline Prices, Apr. 26, 2005, at [http://www.whitehouse.gov/news/releases/2006/04/ 20060425-2.html].

[85] DOE, Factsheet on a Scientific Roadmap for Cellulosic Ethanol, p. 1. Assuming that the 60 billion gallons per year is provided by ethanol, that would be equal to 3.9 million barrels per day of ethanol. Using the fact that

ethanol has about 67% of the heat content of gasoline by volume yields an estimate of 2.6 million barrels of oil per day. See [http://www.er.doe.gov/News_Information/News_Room/2006/Biofuels/factsheet.htm].

[86] DOE, Factsheet on a Scientific Roadmap for Cellulosic Ethanol, p. 2.

[87] DOE, EIA, Annual Energy Outlook 2006, Issues in Focus, p. 58.

[88] About 90% of U.S.biodiesel is made from soybean oil.

[89] Air toxics are defined as any air pollutant for which a national ambient air quality standard (NAAQS) does not exist (i.e., excluding ozone, carbon monoxide, PM-10, sulfur dioxide, nitrogen oxide) that may reasonably be anticipated to cause cancer; respiratory, cardiovascular, or developmental effects; reproductive dysfunctions; neurological disorders; heritable gene mutations; or other serious or irreversible chronic or acute health effects in humans.

[90] DOE, EERE, Alternative Fuels Data Center, Biodiesel, at [http://www.eere.energy.gov/afdc/altfuel/biodiesel.html].

[91] EIA, Biodiesel Performance, Costs, and Use, 2004, 8 p., at [http://www.eia.doe.gov/oiaf/analysispaper/biodiesel/].

[92] DOE, National Renewable Energy Laboratory, An Overview of Biodiesel and Petroleum Diesel Lifecycles, p. v., at [http://www.biodiesel.com/Biodiesel%20Life%20Cycle.pdf]; also see National Biodiesel Board, Energy Content, Fact Sheet, 1 p., at [http://www.biodiesel.org/pdf_files/fuelfactsheets/BTU_Content_Final_Oct2005.pdf].

[93] National Biodiesel Board (NBB), U.S. Biodiesel Production Capacity, May 2006, 1 p.

[94] National Biodiesel Board (NBB), Biodiesel and Energy Security, Fact Sheet, 2 p. NBB reports that on-road diesel fuel use stands at about 37 billion gallons per year; see [http://www.biodiesel.org/pdf_files/fuelfactsheets/Energy_Security.pdf].

[95] EIA, Annual Energy Outlook 2006, p. 52. This is the estimate for the reference case.

[96] The $1.00 per gallon credit is for biodiesel derived from first-use oils, such as soybean oil. For biodiesel obtained from second-use oils, such as recycled cooking oil, the credit is set at half that value.

[97] Thus, the maximum credit for biodiesel derived from second-use oils is 50 cents per gallon for B 100.

[98] However, the share of RFS allocated to biodiesel and other details are to be determined through a rulemaking by the Environmental Protection Agency due August 8, 2006. Meanwhile, EPA has posted the default rule at [http://www.epa.gov/otaq/ renewablefuels/420f05057.htm].

[99] EPACT has other provisions supporting biodiesel, including an authorization for an engine-testing program (§757), a 10-cent per gallon tax credit for small agri-biodiesel producers (§1345), and a requirement that DOE prepare a report on the future potential for biodiesel fuel (§1823).

[100] Because renewable energy is often developed for energy security, air pollution reduction, or other purposes, it is an example of a "no-regrets" strategy for CO2 emission reductions. Wind and solar energy have zero CO2 emissions in operation but may need an energy storage back-up system (such as batteries or fuel cells) that do require fossil fuel use. When biomass is developed as an energy crop, the CO2 emissions are near zero because each new crop absorbs the same amount of emissions as are released by combusting the previous crop.

[101] The report is available at [http://www.climatetechnology.gov/library/2005/tech-options/index .htm].

[102] U.S. Department of State, U.S. Climate Action Report: The United States of America's Third National Communication Under the United Nations Framework Convention on Climate Change, May 2002. The report is available online at [http://yosemite.epa.gov/oar/globalwarming.nsf/content/ResourceCenterPublicationsUSClimateActionReport.html].

[103] United Nations Environment Program (UNEP), Climate Change 2001: Working Group III, Mitigation, at [http://www.grida.no/climate/ipcc_tar/wg3/index.htm].

[104] These laws and bills are described in CRS Report RL32860, Energy Efficiency and Renewable Energy Legislation in the 109th Congress, by Fred Sissine.

Congressional Hearings, Reports, and Documents

U.S. Congress. Senate. Committee on Energy and Natural Resources. United States — India Energy Cooperation. Hearing held July 18, 2006. [http ://energy.senate.gov/public/index.cfm?FuseAction=Hearings.Hearing& Hearing_ID= 1578]

U.S. Congress. Senate. Committee on Energy and Natural Resources. Geothermal Energy. Hearing held July 11, 2006.[http ://energy.senate.gov/public/index.cfm?FuseAction=Hearings .Hearing and Hearing_ID= 1576]

U.S. Congress. Senate. Committee on Energy and Natural Resources. Energy Efficiency (S. 2747). Hearing held June 22, 2006. [http ://energy. senate. gov/public/index.cfm?FuseAction=Hearings .Hearing& Hearing_ID= 1568]

U.S. Congress. Senate. Committee on Energy and Natural Resources. Renewable Fuel Standards. Hearing held June 19, 2006.[http ://energy. senate.gov/ public/index.cfm?FuseAction=Hearings .Hearing and Hearing_ID= 1565]

U.S. Congress. Senate. Committee on Environment. Global Climate Change. Hearing held May 25, 2006. [Not available on Committee website.]

U.S. Congress. Senate. Committee on Commerce, Science, and Transportation. Corporate Average Fuel Economy (CAFE) Standards. Hearing held May 9, 2006. [http ://commerce.senate.gov/public/index.cfm?FuseAction=Hearings. Hearing& Hearing_ID= 1756]

U.S. Congress. Senate. Committee on Energy and Natural Resources. Climate Change. Conference held April 4, 2006. (S. Hrng. 109-420) [http://frwebgate.access.gpo.gov/cgi-bin/getdoc.cgi?dbname=109_senate_ hearings&docid=f:28095 .pdf]

U.S. Congress. Senate. Committee on Energy and Natural Resources. Power Generation Resource Incentives and Diversity Standards. Hearing held March 8, 2005. [http://energy.senate.gov/hearings/witnesslist.cfm?id= 1403]

(A more extensive list appears in CRS Report RL32860, Energy Efficiency and Renewable Energy Legislation in the 109th Congress, by Fred Sissine.)

For Additional Reading

Tables showing DOE Renewable Energy R&D Funding trends back to FY1978 (current and constant) are available from the author of this report.

American Solar Energy Society. Renewable Hydrogen Forum. April 10-11, 2003.

Barry, Courtney. "Winds of Change in Texas." Public Utilities Fortnightly, v. 141, no. 7, April 1, 2003. pp. 27-31.

Blankinship, Steve. "A Sunny Outlook for Grid-Connected PV." Power Engineering, v. 107, no. 1, May 2003. pp. 32-40.

Cato Institute. Policy Analysis. Evaluating the Case for Renewable Energy: Is Government Support Warranted? 2002.

Energy Future Coalition. Challenge and Opportunity: Charting a New Energy Future, Report of the Bioenergy and Agriculture Working Group. 2003. [http://www.energyfuturecoalition.org/]

European Commission. Directorate General for Research. European Research Spending for Renewable Energy Sources. 2004. (European Union) [http://ec.europa.eu/research/energy/pdf/res_spending_en.pdf]

International Energy Agency. Renewables Information 2005. September 2005. [http://library.iea.org/w/bookshop/add.aspx?id=68]

National Research Council. The Hydrogen Economy: Opportunities, Costs, Barriers, and R&D Needs. 2004.

Owens, Brandon. "Does the PTC Work?" Standard and Poor's. September 2004.

Popular Science. "Wind Power Reconsidered." November 2004. pp. 43-44. [http://www.popsci.com/popsci/technology/generaltechnology/72959aa1 38b]

Schimmoller, Brian K. "Renewables Get Into the Mix." Power Engineering, January 2004. pp. 22-30.

Sklar, Scott and Sheinkopf, Kenneth. Consumer Guide to Solar Energy: New Ways to Lower Utility Costs, Cut Taxes, and Take Control of Your Energy Needs. 2002.

The White House. State of the Union. The Advanced Energy Initiative. January 31, 2006. [http://www.whitehouse.gov/news/releases/2006/0 1/20060131 - 6.html]

U.S. Department of Energy. Interlaboratory Working Group. Scenarios for a Clean Energy Future. 2000. [http://www.ornl.gov/sci/eere/cef/]

——. Energy Information Administration. Federal Financial Interventions and Subsidies in Energy Markets 1999: Primary Energy. (SR/OIAF/99-03). 1999.

——. Federal Energy Regulatory Commission. The Interconnection of Wind Energy and Other Alternative Technologies. January 24, 2005. [http://www.ferc.gov/industries/electric/indus-act/gi/wind.asp].

——. Lawrence Berkeley National Laboratory. Balancing Cost and Risk: The Treatment of Renewable Energy in Western Utility Resource Plans. [Mark Bolinger and Ryan Wiser] August 2005. [http://eetd.lbl.gov/ea/ems/re-pubs.html]

——. National Renewable Energy Laboratory. Near-Term Practical and Ultimate Technical Potential for Renewable Resources. [Draft] January 16, 2006. 7 p.

——. National Renewable Energy Laboratory. Domestic Energy Scenarios. 2003. [http://www.nrel.gov/docs/fy03osti/32742.pdf]

——. National Renewable Energy Laboratory. Projected Benefits of Federal Energy Efficiency and Renewable Energy Programs: FY2006 Budget Request. May 2005. [http://www.nrel.gov/docs/fy05osti/3793 1 .pdf]

——. National Renewable Energy Laboratory. Power Technologies Energy Data Book. April 2005. [http://www.nrel.gov/docs/fy03osti/32742.pdf]

—— Oak Ridge National Laboratory. Biomass as Feedstock for a Bioenergy and Bioproducts Industry: The Technical Feasibility of a Billion-Ton Annual Supply. [DOE/USDA Biomass Feedstock Gate Review Meeting] March 2005. [http://feedstockreview.ornl.gov/pdf/perlack_stokes/session1_perlack_stokes .pdf]

——. Office of Science. Basic Research Needs for Solar Energy Utilization. August 2005. [http://www.sc.doe.gov/bes/reports/files/SEU_rpt.pdf]
——. Enhancing Homeland Security Through Renewable Energy — Richard Truly's Remarks to the National Press Club. March 14, 2002.
——. Status Report to Congress on Current and Proposed Activities under the Clean Energy Technology Exports (CETE) Initiative. 2001. [http://www.pi.energy.gov/library/cete200 1 statusreport.html#pdf]
——. The Clean Air Act and Renewable Energy: Opportunities, Barriers, and Options. (NREL/CP-620-29654). 2001.
U.S. Department of the Interior. Bureau of Land Management. Solar Energy Development Policy. [Instruction Memo No. 2005-006] October 20, 2004. [http://www.blm.gov/nhp/efoia/wo/fy05/im2005-006.htm]
U.S. Environmental Protection Agency. Climate Action Report: The United States of America's Third National Communication under the United Nations Framework Convention on Climate Change. 2002. [http://yosemite. epa.gov/oar/globalwarming.nsf/content/ResourceCenterPublicationsUSClimat eActionReport.html]
U.S. Executive Office of the President. Office of Science and Technology Policy. Domestic Policy Council. American Competitiveness Initiative. February 2006. 23 p. [http://www.ostp.gov/html/ACIBooklet.pdf]
U.S. Government Accountability Office (GAO). Renewable Energy: Wind Power's Contribution to Electric Power Generation and Impact on Farms and Rural Communities. [GAO-04-756] 2004. [http://www.gao.gov/docsearch/ abstract.php?rptno=GAO-04-756]
Worldwatch Institute. Mainstreaming Renewable Energy in the 21st Century. May 2004. [http://www.worldwatch.org/pubs/paper/169/]

CRS Reports

CRS Report RL327 12. Agriculture-Based Renewable Energy Production, by Randy Schnepf.
CRS Report RL33564. Alternative Fuels and Advanced Technology Vehicles: Issues in Congress, by Brent Yacobucci.
CRS Report RL33523. Arctic National Wildlife Refuge (ANWR): Controversies for the 109th Congress, by M. Lynne Corn, Bernard A. Gelb, and Pamela Baldwin.
CRS Report RS2 1563. Biodiesel Fuel and U.S. Agriculture, by Randy Schnepf.

CRS Report RL33294. DOE Budget Earmarks: A Selective Look at Energy Efficiency and Renewable Energy R&D Programs, by Fred Sissine.

CRS Report RL3 1033. Energy Efficiency and Renewable Energy Fuel Equivalents to Potential Oil Production from the Arctic National Wildlife Refuge (ANWR), by Fred Sissine.

CRS Report RL3 2860. Energy Efficiency and Renewable Energy Legislation in the 109th Congress, by Fred Sis sine.

CRS Report RL33578. Energy Tax Policy: History and Current Issues, by Salvatore Lazzari.

CRS Report RL33290. Fuel Ethanol: Background and Public Policy Issues, by Brent Yacobucci.

CRS Report RS2 1442. Hydrogen and Fuel Cell Vehicle R&D: freedomCAR and the President's Hydrogen Fuel Initiative, by Brent Yacobucci.

CRS Report RL32865. Renewable Fuels and MTBE: A Comparison of Selected Legislative Initiatives, by Brent Yacobucci, Mary Tiemann, and James McCarthy.

Websites

American Council for Renewable Energy. [http://www.acore.org/]

American Solar Energy Society. [http://www.ases.org/]

American Wind Energy Association (AWEA). [http://www.awea.org/]

California Energy Commission. [http://www.energy.ca.gov/renewables/index.html]

Database of State Incentives for Renewable Energy (IREC). [http://www.dsireusa.org/]

Databook of Renewable Energy Power Technologies. [http://www.nrel.gov/analysis/power_databook/]

Earthtrack (Database of Energy Subsidies). [http://www.earthtrack.net/]

Edison Electric Institute. [http://www.eei.org/]

Electric Power Research Institute (EPRI) and EPRI Journal Online. [http://www.epri.com/]

Eleventh Session of the Conference of Parties to the United Nations Framework Convention on Climate Change (Montreal, COP-1 1 and COP/MOP 1). November 28 - December 9, 2005. [http://unfccc.int/meetings/cop_1 1/items/3394.php]

National Association of Regulatory Utility Commissioners. [http://www.naruc.org/]

Renewable Energy Policy Project. [http://www.crest.org/]
Renewable Energy/Information Center. International Energy Agency (IEA). [http://www.iea.org/dbtw-wpd/Textbase/subjectqueries/index.asp]
Solar Electric Power Association (SEPA). [http://www.solarelectricpower.org/]
Solar Energy Industries Association (SEIA). [http://www.seia.org/]
Tax Incentives Assistance Project. [http://www.energytaxincentives.org/]
U.S. Department of Energy. Energy Efficiency and Renewable Energy Network. [http://www.eere.energy.gov/]
U.S. Department of Energy. Green Power Network Clearinghouse. [http://www.eere.energy.gov/greenpower/]
U.S. Department of Energy. National Renewable Energy Laboratory (NREL). [http://www.nrel.gov/]
U.S. Department of Energy. Alternative Fuels Data Center. [http://www.eere.energy.gov/afdc/index.html]
U.S. Environmental Protection Agency. Clean Energy Site. [http://www.epa.gov/cleanenergy/]
Vote Solar Initiative. San Francisco's $100 Million Solar Revenue Bond Initiative. [http://www.votesolar.org/sf.html].

In: Renewable Fuel Standard Issues
Editor: Daniel T. Crowe

ISBN 978-1-60692-289-7

Chapter 3

ENERGY INDEPENDENCE AND SECURITY ACT OF 2007: A SUMMARY OF MAJOR PROVISIONS*

Fred Sissine

ABSTRACT

The Energy Independence and Security Act (P.L. 110-140, H.R. 6) is an omnibus energy policy law that consists mainly of provisions designed to increase energy efficiency and the availability of renewable energy. This report describes the key provisions of the enacted law, summarizes the legislative action on H.R. 6, and provides a summary of the provisions under each of the titles in the law.

The highlights of key provisions enacted into law are as follows:

- Corporate Average Fuel Economy (CAFE). The law sets a target of 35 miles per gallon for the combined fleet of cars and light trucks by model year 2020.
- Renewable Fuels Standard (RFS). The law sets a modified standard that starts at 9.0 billion gallons in 2008 and rises to 36 billion gallons by 2022.
- Energy Efficiency Equipment Standards. The adopted bill includes a variety of new standards for lighting and for residential and commercial appliance equipment. The equipment includes residential refrigerators,

* Excerpted from CRS Report RL34294, dated December 21, 2007.

freezers, refrigerator-freezers, metal halide lamps, and commercial walk-in coolers and freezers.

- Repeal of Oil and Gas Tax Incentives. The enacted law includes repeal of two tax subsidies in order to offset the estimated cost to implement the CAFE provision.

The two most controversial provisions of H.R. 6 that were not included in the enacted law were the proposed Renewable Energy Portfolio Standard (RPS) and most of the proposed tax provisions, which included repeal of tax subsidies for oil and gas and new incentives for energy efficiency and renewable energy.

Introduction

The Energy Independence and Security Act (P.L. 110-140, H.R. 6) is an omnibus energy policy law that consists mainly of provisions designed to increase energy efficiency and the availability of renewable energy. This report describes the key provisions of the enacted law, summarizes the legislative action on H.R. 6, and provides a summary of the provisions under each of the titles in the law.

Many analysts in the CRS Resources, Science, and Industry Division contributed to this article; their names and contact information are located on the back of the summary page.

Key Provisions

Provisions Included

The three key provisions enacted in P.L. 110-140 are the Corporate Average Fuel Economy (CAFE) Standards, the Renewable Fuel Standard (RFS), and the appliance/lighting efficiency standards.

Corporate Average Fuel Economy (CAFE) Standards. The law sets a target of 35 miles per gallon for the combined fleet of cars and light trucks by model year 2020. Also, a fuel economy program is established for medium- and heavy-duty trucks, and a separate fuel economy standard is created for work trucks. (For more details on issues related to the CAFE provision, see CRS Report RL33982, Corporate Average Fuel Economy (CAFE): A Comparison of Selected Legislation in the 110th Congress, by Brent D. Yacobucci and Robert Bamberger, and CRS Report RL334 13, Automobile and Light Truck Fuel Economy: The CAFE Standards, by Brent D. Yacobucci and Robert Bamberger.)

Renewable Fuel Standard (RFS). The law sets a modified standard that starts at 9.0 billion gallons of renewable fuel in 2008 and rises to 36 billion gallons by 2022. Of the latter total, 21 billion gallons is required to be obtained from cellulosic ethanol and other advanced biofuels. (For more details on issues related to the RFS provision, see CRS Report RL34265, Selected Issues Related to an Expansion of the Renewable Fuel Standard (RFS).)

Appliance and Lighting Efficiency Standards. Energy efficiency standards are set for broad categories of incandescent lamps (light bulbs), incandescent reflector lamps, and fluorescent lamps. A required target is set for lighting efficiency, and energy efficiency labeling is required for consumer electronic products. Also, efficiency standards are set by law for external power supplies, residential clothes washers, dishwashers, dehumidifiers, refrigerators, refrigerator- freezers, freezers, electric motors, residential boilers, commercial walk-in coolers, and commercial walk-in freezers. Further, DOE is directed to set standards by rulemaking for furnace fans and battery chargers.

Provisions Excluded

Two controversial provisions of H.R. 6 that were not included in the enacted law were the proposed Renewable Energy Portfolio Standard (RPS) and the proposed repeal of tax subsidies for oil and gas.

Renewable Energy Portfolio Standard (RPS). Under an RPS, retail electricity suppliers (electric utilities) must provide a minimum amount of electricity from renewable energy resources or purchase tradable credits that represent an equivalent amount of renewable energy production. The minimum requirement is often set as a percentage share of a supplier's total retail electricity sales. The second degree amendment to H.R. 6 passed by the House on December 6, 2007, proposed a national RPS target that aimed to reach 15% of total electricity sales by 2020. Up to 4 percentage points of the 15% target could be met with energy efficiency measures. This provision was stripped out by the Senate and was not included in the final version of the bill. (For more details on issues related to the RPS provision, see CRS Report RL341 16, Renewable Energy Portfolio Standard (RPS): Background and Debate Over a National Requirement, by Fred Sissine.)

Energy Tax Subsidies. The House-passed second degree amendment to H.R. 6 contained provisions that would have repealed about $22 billion of oil and gas subsidies that were designed to offset the cost of supporting a variety of energy efficiency and renewable energy tax incentives. These proposed incentives would have included a four-year extension of the renewable energy electricity production tax credit. Most of those provisions were stripped out by the Senate and were not included in the final bill. Enough tax revenue offsets were included to cover the estimated cost of the CAFE provision. (For more details about the proposed renewable energy incentives, see CRS Report RL34 162, Renewable Energy Issues in the 110th Congress, by Fred Sissine. For more details about the proposed repeal of oil and gas subsidies, see CRS Report RL33578, Energy Tax Policy, by Salvatore Lazzari.)

BRIEF SUMMARY OF LEGISLATIVE ACTION

On January 18, 2007, the House passed the 14-page CLEAN Energy Act (H.R. 6) by a vote of 264-163. The bill proposed to repeal certain oil and natural gas subsidies, thus generating revenue for an Energy Efficiency and Renewables Reserve. The Reserve was designed to reduce foreign oil dependence and serve other purposes. The actual uses of the reserve would have been determined at a later date by further legislation. [1]

On June 21, 2007, the Senate passed its first degree amendments to H.R. 6 as the proposed Renewable Fuels, Consumer Protection, and Energy Efficiency Act of 2007. This action transformed H.R. 6 into a 500-page omnibus energy policy bill, with a primary focus on energy efficiency and renewable energy. The Senate's amendments were drawn primarily from the proposed Energy Savings Act of 2007 (S. 1321). [2] The key provisions of the Senate-passed H.R. 6 were appliance efficiency standards, an increase of the renewable fuel standard to 36 billion gallons by 2022, and an increase of the combined corporate average fuel economy standards to 35 miles per gallon (mpg) by 2020.

On August 4, 2007, the House passed the omnibus energy policy bill, H.R. 3221, which had two divisions and 13 titles. Division A contained provisions of the New Direction for Energy Independence, National Security, and Consumer Protection Act. An adopted floor amendment (H.Amdt. 748) added a 15% renewable energy portfolio standard (RPS). Division B, the Renewable Energy and Energy Conservation Tax Act of 2007, was also added as a floor amendment and contained the House-approved version of H.R. 2776. It added four titles to H.R. 3221 that included a four-year extension of the renewable electricity production tax credit and other efficiency and renewables incentives.

Because the House omnibus bill (H.R. 3221) and the Senate omnibus bill (H.R. 6) had different bill numbers, the bills could not be taken directly to conference committee. [3] However, after the House completed action on H.R. 3221, informal bipartisan negotiations over the omnibus energy bills began between the House and Senate. Key issues included CAFE, the renewable fuel standard, the RPS provision in H.R. 3221, and a proposed repeal of certain oil and natural gas subsidies to offset costs for new energy efficiency and renewable energy tax incentives.

On December 6, 2007, the House passed by a vote of 235-181 its second-degree amendments to the Senate-passed amendments to H.R. 6. This "second version" of a House omnibus energy bill was derived primarily by trimming and modifying H.R. 3221 and adding major new provisions on CAFE and RFS. The House-passed bill included a proposed increase of the CAFE standard to 35 miles per gallon by 2020 and an increase of the renewable fuel standard to 36 billion gallons per year by 2022. The House bill also included a proposed 15% renewable electricity portfolio standard and $21 billion of new tax incentives for energy efficiency and renewable energy measures. The bill proposed to offset the new tax incentives with a repeal of about $21 billion in tax subsidies for oil and natural gas.

The White House threatened to veto the House-passed bill mainly because of the presence of provisions for an RPS and for the repeal of oil and gas tax subsidies. [4] On December 7, 2007, a Senate cloture vote on the House-passed version of H.R. 6 — with provisions for an RPS and for the repeal of oil and gas subsidies — failed (5 2-43). After stripping out the RPS and modifying the tax package, a cloture vote on S.Amdt. 3841 failed (59-40).

On December 13, 2007, the Senate adopted S.Amdt. 3850, its second-degree substitute amendment to H.R. 6 by a vote of 86-8. The Senate substitute was nearly identical to the House-passed bill, except that the RPS provision and most tax provisions had been taken out. The House approved the Senate bill on December 18, 2007. The President signed the bill into law on December 19, 2007 (P.L. 110-140).

Title I: Energy Security Through Improved Vehicle Fuel Economy

Subtitle A, Increased Corporate Average Fuel Economy

This subtitle requires an increase in CAFE standards and a restructuring of the fuel economy program. A single CAFE standard of 35 miles per gallon (mpg) by

MY2020 is established, and the distinction between the passenger car and light truck fleet is preserved. The new standards will be based on vehicle attributes and expressed in the form of a mathematical function. Interim standards will be set, beginning with MY201 1. Manufacturers will be required to come within 92% of the standard for a given model year. However, manufacturers can earn credits for exceeding the standards in one vehicle class that can be applied to boost, within limitations, the CAFE of a different vehicle class that is falling short of compliance. Additionally, credits may be sold and bought between manufacturers. CAFE credits for the manufacture of flexible-fueled vehicles (FFV) are retained but phased out by MY2020. Civil penalties assessed for non-compliance will be deposited to the general fund of the U.S. Treasury to support future rulemaking and to provide grants to manufacturers for research and development, and retooling in support of increasing the fuel efficiency of their fleets. The law requires the development of standards for "work trucks" and commercial medium- and heavy-duty on-highway vehicles. (For additional information, see CRS Report RL33413, Automobile and Light Truck Fuel Economy: The CAFE Standards, by Brent D. Yacobucci and Robert Bamberger.)

Subtitle B, Improved Vehicle Technology

This subtitle establishes a loan guarantee program for advanced battery development, grant programs for plug-in hybrid vehicles, incentives for purchasing heavy-duty hybrid vehicles for fleets, and credits for various electric vehicles.

Subtitle C, Federal Vehicle Fleets

Federal agencies are prohibited from acquiring any light-duty motor vehicle or medium-duty passenger vehicle that is not "a low greenhouse gas emitting vehicle" as defined in this subtitle. Alternatively, the agency may demonstrate that it has adopted cost-effective policies to reduce its petroleum consumption sufficiently to achieve a comparable reduction in greenhouse gas emissions. By 2015, federal agencies are required to achieve at least a 20% reduction in annual petroleum consumption and a 10% increase in annual alternative fuel consumption. These increases are to be calculated from a 2005 baseline. Interim milestones will be established and agencies will report annually on their progress.

The regulations governing this program are required to be issued not later than 18 months after enactment.

TITLE II: ENERGY SECURITY THROUGH INCREASED PRODUCTION OF BIOFUELS

Subtitle A, Renewable Fuel Standard

This subtitle extends and increases the renewable fuel standard (RFS) set by P.L. 109-58 (§1501). The RFS requires minimum annual levels of renewable fuel in U.S. transportation fuel. The previous standard was 5.4 billion gallons for 2008, rising to 7.5 billion by 2012. The new standard starts at 9.0 billion gallons in 2008 and rises to 36 billion gallons in 2022. Starting in 2016, all of the increase in the RFS target must be met with advanced biofuels, defined as cellulosic ethanol and other biofuels derived from feedstock other than corn starch — with explicit carve-outs for cellulosic biofuels and biomass-based diesel. The EPA Administrator is given authority to temporarily waive part of the biofuels mandate, if it were determined that a significant renewable feedstock disruption or other market circumstance might occur. Renewable fuels produced from new biorefineries will be required to reduce by at least 20% the life cycle greenhouse gas (GHG) emissions relative to life cycle emissions from gasoline and diesel. Fuels produced from biorefineries that displace more than 80% of the fossil-derived processing fuels used to operate a biofuel production facility will qualify for cash awards. Several studies are required on the impacts of an RFS expansion on various sectors of the economy. (For more details on issues related to the RFS proposal, see CRS Report RL34265, Selected Issues Related to an Expansion of the Renewable Fuel Standard (RFS), by Brent D. Yacobucci and Randy Schnepf.)

Subtitle B, Biofuels Research and Development (R&D)

This subtitle promotes research on the expansion of the use of biodiesel and biogas as motor fuels. Grants are authorized for R&D and commercial applications of cellulosic biofuels technologies and for the conversion of existing corn-based ethanol plants to produce cellulosic biofuels. The Secretary of Energy is required to report to Congress on the feasibility of algae as a feedstock for biofuels production. The subtitle also promotes university-based R&D on biofuels.

Subtitle C, Biofuels Infrastructure

This subtitle aims to improve information about federal biofuels research programs, focus research on infrastructure and biorefineries, study potential impacts of increased biofuels use, and increase authorized funding for DOE biofuels research. A funding authorization of $25 million is established to provide grants for biofuels research, development, and demonstration (RD&D) and commercial applications in states that have low rates of ethanol production. A university-based program is authorized to provide grants of up to $2 million for R&D on renewable energy technologies. Priority is given to universities in low-income and rural communities with proximity to trees dying of disease or insect infestation.

DOE is directed to create a grant program to help establish or convert infrastructure to use renewable fuels, including E85 (85% ethanol). The Energy Policy Act of 2005 (EPACT, P.L. 109-58) authorization for grants to support cellulosic ethanol production is increased. A grant program is authorized to support production of flexible-fueled vehicles. Studies are also required on the market penetration of flexible-fueled vehicles, the feasibility of constructing dedicated ethanol pipelines, the feasibility of using greater percentages of ethanol in fuel blends, and the adequacy of railroad transportation for delivery of ethanol fuel.

Subtitle D, Environmental Safeguards

Previously, under the Clean Air Act (§21 1 [f]), no new fuels or fuel additives could be introduced into commerce unless granted a waiver by the Environmental Protection Agency (EPA). If EPA did not act within 180 days of receiving a waiver request, the waiver was treated as granted. [5] Section 251 tightens the waiver provision. It amends the Clean Air Act to prohibit the introduction of new fuels or fuel additives unless EPA explicitly grants a waiver. After receiving a waiver request, EPA will now have 270 days to take final action.

TITLE III: ENERGY SAVINGS THROUGH IMPROVED STANDARDS FOR APPLIANCES AND LIGHTING

Subtitle A, Appliance Energy Efficiency

This title sets, by statute, new efficiency standards for external power supplies, residential clothes washers, dishwashers, dehumidifiers, refrigerators, refrigerator-freezers, freezers, electric motors, and residential boilers. DOE is allowed to establish regional variations in standards for heating and air conditioning equipment. DOE is required to complete a rulemaking process for furnace fans by 2013. Federal agencies are directed to purchase devices that limit standby power use. DOE is directed to issue a final rule that sets efficiency standards for battery chargers. Certain energy efficiency measures for walk-in coolers and walk-in freezers are set by law. Also, several procedural changes are now in place to expedite the DOE rulemaking process.

Subtitle B, Lighting Energy Efficiency

Section 321 sets an energy efficiency standard for general service incandescent lamps, provides for consumer education and lamp labeling, and requires market assessments and a consumer awareness program. Section 322 sets energy efficiency standards for incandescent reflector lamps and fluorescent lamps. For federal buildings, Section 323 sets energy efficiency requirements for GSA-leased space and for use of energy efficient lighting fixtures and bulbs in those leased spaces. Section 324 sets energy efficiency standards for metal halide lamp fixtures designed to be operated with lamps rated between 150 watts and 500 watts. Section 325 directs the Consumer Product Safety Commission to set energy efficiency labeling requirements for consumer electronic products.

TITLE IV: ENERGY SAVINGS IN BUILDINGS AND INDUSTRY

Subtitle A, Residential Building Efficiency

Section 411 increases the funding authorization for DOE's Weatherization Program, providing \$3.75 billion over five years. Under Section 412, DOE is directed to conduct a study of the renewable energy system rebate program described in §206(c) of the Energy Policy Act of 2005. The study aims to determine the minimum funding the program would need to be viable. Further, DOE is directed to propose an implementation plan. Section 413 requires DOE to establish energy efficiency standards for manufactured housing.

Subtitle B, High-Performance Commercial Buildings

This subtitle encourages the development of more energy-efficient "green" commercial buildings. Section 421 creates an Office of Commercial High Performance Green Buildings at DOE. Section 422 establishes a zero-energy commercial buildings initiative. A national goal is set to achieve zero-net-energy use for new commercial buildings built after 2025. A further goal is to retrofit all pre- 2025 buildings to zero-net-energy use by 2050. Section 423 requires that DOE establish a national clearinghouse for information and public outreach about high- performance green buildings.

Subtitle C, High-Performance Federal Buildings

Section 431 requires that total energy use in federal buildings, relative to the 2005 level, be reduced 30% by 2015. Section 432 directs that federal energy managers conduct a comprehensive energy and water evaluation for each facility at least once every four years. For new federal buildings and major renovations, Section 433 requires that fossil-fuel energy use — relative to the 2003 level — be reduced 55% by 2010 and be eliminated (100% reduction) by 2030. Section 434 requires that each federal agency ensure that major replacements of installed equipment (such as heating and cooling systems), or renovation or expansion of existing space, employ the most energy efficient designs, systems, equipment, and controls that are life-cycle cost effective. Section 435 prohibits federal agencies from leasing buildings that have not earned an EPA Energy Star label. Section 436 requires GSA to establish an Office of Federal High-Performance Green Buildings to coordinate green building information and activities within GSA and with other federal agencies. The Office must also develop standards for federal facilities, establish green practices, review budget and life-cycle costing issues, and promote demonstration of innovative technologies. Section 437 directs the Government Accountability Office (GAO) to audit the implementation of activities required under this subtitle. The audit must cover budget, life-cycle costing, contracting, best practices, and agency coordination. Section 438 requires federal facility development projects with a footprint exceeding 5,000 square feet to use site planning, design, construction, and maintenance strategies to control storm water runoff. Section 439 directs GSA to review the current use of, and design a strategy for increased use of, cost-effective lighting, ground source heat pumps, and other technologies in GSA facilities. Section 440 authorizes $4 million per year over five years to support work under sections 434-439 and 482. For the purpose of conducting life-cycle cost calculations, Section 441 increases the time period from 25 years, in prior law, to 40 years.

Subtitle D, Industrial Energy Efficiency

Section 451 directs DOE to conduct research on, develop, and demonstrate new processes, technologies, and operating practices and techniques to significantly improve the energy efficiency of equipment and processes used by energy-intensive industries. Section 452 directs EPA to establish a recoverable waste energy inventory program. This program must include an ongoing survey of all major industrial and large commercial combustion sources in the United States. EPA is required to identify the potential for economically feasible waste energy recovery, create a grant program to support waste energy recovery, and strengthen "clean energy centers" that analyze waste energy recovery. Section 453 directs DOE to initiate a voluntary national information program for widely used data centers and data center equipment for which there is significant potential for energy savings. DOE is also tasked with helping to devise strategies to improve energy efficiency at these data centers.

Subtitle E, Healthy High-Performance Schools

Section 461 creates a grant program for Healthy High-Performance Schools that aims to encourage states, local governments, and school systems to build green schools. EPA, in consultation with the Department of Education, is allowed to provide grants to state agencies to provide technical assistance and help with the development of state plans for school building design. Also, EPA is directed to develop model voluntary guidelines for school site selection. In addition to other environmental aspects, the grants and guidelines must have a focus on energy efficiency, natural daylighting, and other energy-related features. Section 462 directs EPA to lead a detailed study of how sustainable building features, such as energy efficiency, affect multiple perceived indoor environmental quality stressors on students in K-12 schools.

Subtitle F, Institutional Entities

Section 471 creates a program of grants and loans to support energy efficiency and energy sustainability projects at public institutions.

Subtitle G, Public and Assisted Housing

Section 481 directs the Department of Housing and Urban Development (HUD) to update energy efficiency standards for all public and assisted housing.

Subtitle H, General Provisions

Section 491 calls for the DOE Office of Commercial High Performance Buildings and the GSA Office of Federal High Performance Buildings to jointly develop guidelines for demonstration projects. In accordance with the guidelines, one federal project must be undertaken annually over a five-year period, supported by a $10 million funding authorization. Also, a total of four projects are to be undertaken at different universities over the five-year period, supported by an additional $10 million funding authorization. Section 492 calls for these two offices to undertake a joint survey of research on green buildings, coordinate efforts to develop a research plan, and identify potential benefits of green buildings for security, natural disasters, and emergency needs of the federal government. Section 493 requires EPA to create a program of competitive grants to local governments for green building demonstration projects. Section 494 directs the Office of Commercial High Performance Buildings and the Office of Federal High Performance Buildings to jointly appoint a Green Building Advisory Committee with representatives from a variety of backgrounds, including federal agencies, state and local governments, building industry experts, security advisors, and environmental health experts. Section 495 calls for DOE to create an advisory committee on energy efficiency finance to help lower costs and increase investment for energy efficiency technologies.

TITLE V: ENERGY SAVINGS IN GOVERNMENT AND PUBLIC INSTITUTIONS

Subtitle A, United States Capitol Complex

Section 501 allows the Architect of the Capitol (AOC) to perform a feasibility study regarding construction of a photovoltaic roof for the Rayburn House Office Building. Under Section 502, the AOC is allowed to construct a fuel tank and pumping system for E85 (85% ethanol) fuel at or within close proximity to the Capitol Grounds Fuel Station. Section 503 requires the AOC, to the maximum extent practicable, to include energy efficiency measures, climate change mitigation measures, and other appropriate environmental measures in the Capitol Complex Master Plan. Under Section 504, the AOC is directed to operate the steam boilers and chiller plant at the Capitol Power Plant in the most energy efficient manner possible to minimize carbon emissions and operating costs. Further, Section 505 requires the AOC to install technologies for the capture and storage or use of carbon dioxide emitted from coal combustion in the Capitol Power Plant.

Subtitle B, Energy Savings Performance Contracting

Section 511 eliminates the advance reporting requirement for Energy Savings Performance Contracts (ESPCs) that have a cancellation ceiling exceeding $10 million. Section 512 increases ESPC funding flexibility by allowing a combination of appropriated funds and private financing. Section 513 restricts federal agencies from limiting the duration of ESPCs to less than 25 years or limiting the total amount of obligations. Further, this section permits the criteria for savings verification to satisfy the requirement for energy audits. Also, it directs federal agencies to modify existing ESPCs to conform with the requirements of this subtitle. Section 514 permanently authorizes ESPCs.

Section 515 extends the definition of energy savings reduction to include increased use of an existing energy source by cogeneration or heat recovery, use of excess electrical or thermal energy generated from onsite renewable sources or cogeneration, and increased energy-efficient use of water resources. Section 516 permits agencies to retain the full amount of energy and water cost savings obtained from utility incentive programs. Section 517 authorizes $750,000 per year over five years for a program to train contract officers in negotiating ESPCs.

Section 518 directs the Department of Defense (DOD) and DOE to study the potential use of ESPCs in nonbuilding applications, which is defined to include vehicles and federally owned equipment to generate electricity or transport water.

Subtitle C, Energy Efficiency in Federal Agencies

Under Section 521, GSA is directed to use up to $30 million — subject to appropriation — from FY2007 and prior years' unobligated balances of the Federal Buildings Fund to support the installation of a solar photovoltaic system for the DOE headquarters building in the District of Columbia. Section 522 prohibits, except under certain circumstances, the purchase of incandescent light bulbs for use in Coast Guard office buildings. Section 523 requires 30% of the hot water demand in new federal buildings (and major renovations) to be met with solar hot water equipment, provided it is life-cycle cost-effective. Section 524 encourages federal agencies to minimize standby energy use in purchases of energy-using equipment. Section 525 requires federal procurement to focus on use of Energy Star and Federal Energy Management Program (FEMP)-designated products. Section 526 prohibits federal agencies from procuring synfuel unless its life cycle GHG emissions are less than those for conventional petroleum sources. Section 527 directs each federal agency subject to any requirements under this title to issue an annual report that describes the status of initiatives to improve energy efficiency, reduce energy costs, and reduce GHG emissions. Section 528 requires the Office of Management and Budget (OMB) to submit an annual report to Congress that summarizes the information reported under Section 527, evaluates overall progress toward the goals of Section 527, and recommends additional actions needed to meet those goals. Section 529 directs the Federal Energy Regulatory Commission (FERC) to conduct a national assessment of demand response, including an estimate of nationwide demand response out to a 10- year horizon. Further, FERC is required to prepare a National Action Plan on Demand Response, with cooperation from industry. Annual funding of $10 million per year is authorized over three years.

Subtitle D, Energy Efficiency of Public Institutions

Section 531 increases annual funding authorizations for DOE's state energy programs. Under Section 532, electric and natural gas utilities are required to make energy efficiency a priority resource and to integrate energy efficiency into resource plans and planning processes. Further, the utilities are directed to modify their rates to align their incentives with the delivery of cost-effective energy efficiency and promote energy efficiency investments. Utilities are encouraged to consider several policy options for achieving those goals.

Subtitle E, Energy Efficiency and Conservation Block Grants

This subtitle establishes an energy efficiency block grant program. Section 541 provides definitions of program elements. Section 542 directs DOE to establish an energy efficiency and conservation block grant program to help reduce energy use and emissions at the local and regional level. Section 543 establishes allocation percentages for grants provided under this subtitle. Section 544 enumerates the allowed purposes for the use of funds provided under this subtitle, which includes strategic planning, consultant services, and energy audits. Section 545 provides eligibility requirements for grants under this program, including payment of prevailing wage rates, submission of a strategic plan, and sharing of information. Section 546 sets criteria for minimum allocations of competitive grant funding. Section 547 specifies that DOE may review and evaluate the performance of grant recipients and withhold funds from those it deems have failed to achieve compliance. To support the grant program, Section 548 authorizes $2 billion annually over five years. Additional funding is authorized to cover administrative costs of the program. Section 548 stresses that funding will supplement, not replace, funding provided by DOE under the Weatherization and State Energy programs.

TITLE VI: ACCELERATED RESEARCH AND DEVELOPMENT

Subtitle A, Solar Energy

Section 602 aims to improve the cost and effectiveness of thermal energy storage technologies that could improve the operation of concentrating solar power electric generating plants. Section 603 calls for improved integration of concentrating solar power into regional electricity transmission systems.

Subtitle B, Geothermal Energy

DOE is directed to support programs of R&D, demonstration, and commercial application to expand the use of geothermal energy. Section 613 directs DOE to support programs that (1) develop advanced prospecting tools to locate and develop hidden geothermal resources, and (2) demonstrate advanced exploratory drilling technologies and techniques with industry partners. Section 614 directs DOE to support programs to develop components and systems necessary to develop, produce, monitor, and model the performance of geothermal reservoirs used to produce geothermal energy. In addition, Section 614 directs DOE to support programs that mitigate or prevent environmental damage from geothermal energy development.

Section 615 directs DOE to support enhanced geothermal system development, whereby geothermal reservoir systems are engineered (as opposed to naturally occurring systems) by creating fractures and permeable conduits via reservoir stimulation. DOE would support R&D programs for enhanced geothermal system technologies and for reservoir stimulation and support demonstration projects at a minimum of four sites.

DOE is directed to establish a program of R&D, demonstration, and commercial application for geothermal energy production from oil and gas fields and from geopressured resources. [6] Section 616 directs DOE to implement a grant program for at least three demonstration projects that use geothermal techniques to extract energy from marginal, unproductive, and productive oil and gas fields. Also, DOE is directed to establish a grant program for the recovery of energy from geopressured resources.

Section 618 directs DOE to establish a Center for Geothermal Technology Transfer, via a grant to an institution of higher learning or consortium thereof, that would serve as an information clearinghouse for the geothermal industry, make data available to the public, and coordinate R&D efforts among national and international partners. Section 619 would rename DOE's GeoPowering the West program as "GeoPowering America" and expand its geothermal technology transfer activities to cover the entire United States. Section 620 would award a grant on a competitive basis to an institution of higher education to establish a geothermal-powered energy generation facility on the institution's campus.

Section 624 directs DOE to support international geothermal energy development through collaborative efforts to promote geothermal R&D and deployment of geothermal technologies. Section 625 directs DOE to make grants to eligible entities from "high-cost regions" [7] of the United States for a feasibility study, demonstration, and commercial application of technologies related to geothermal energy.

Subtitle B authorizes $90 million annually for geothermal activities, of which $10 million is designated for activities under Section 616. An additional $5 million is authorized annually for the Intermountain West Geothermal Consortium, and $5 million is authorized annually for Section 624. All of the foregoing authorizations are in effect from 2008 to 2012.

Subtitle C, Marine and Hydrokinetic Renewable Energy Technologies

DOE is directed to create an R&D program focused on technology that produces electricity from waves, tides, currents, and ocean thermal differences (§633). A report to Congress is required. Further, DOE is instructed to award grants to institutions of higher education (or consortia thereof) to establish National Marine Renewable Energy Research, Development, and Demonstration Centers (§634).

Subtitle D, Energy Storage for Transportation and Electric Power

The U.S. Energy Storage Competitiveness Act of 2007 directs DOE to conduct a cost-shared RD&D program to support the ability of the nation to remain globally competitive in energy storage systems for electric drive vehicles, stationary applications, and electricity transmission and distribution. An Energy Storage Advisory Council will be created, with responsibility for preparing a five-year research plan. Also, through competitive bids, DOE will establish four energy storage research centers managed by the Office of Science. DOE is required to conduct energy storage demonstration projects. Also, DOE is to investigate secondary applications of energy storage equipment and to examine technologies and processes for final recycling and disposal of energy storage equipment. After five years of program operation, the law will require a review of the program by the National Academy of Sciences. A total authorization of nearly $3 billion is provided over a 10-year period.

Subtitle E, Miscellaneous Provisions

Section 651 directs DOE to establish an RD&D program to determine ways in which the weight of motor vehicles could be reduced to improve fuel efficiency without compromising passenger safety. This will focus on the development of new materials and on reducing the cost of lightweight materials. An $80 million authorization is provided over a five-year period.

Section 652 directs DOE to report on the state of technology development for "advanced" insulation with an R-value greater than R35 per inch. The report is to include an estimate of potential cost savings by applying such insulation to covered refrigeration units. If sufficient cost savings are projected, DOE will then be directed to conduct a cost-shared demonstration program to show actual cost savings. An $8 million funding authorization is provided for that program.

Section 653 changes the sulfur dioxide (SO2) criterion for clean coal power plants from a percentage basis (99% of SO2 removed) to a weight-by-energy basis (no more than 0.04 pounds of SO2 per million Btu).

Section 654 on the "H-Prize" directs DOE to conduct a competitive program to award cash prizes to advance R&D, demonstration, and commercial application of hydrogen energy technologies. Prizes can be a mix of federal appropriations and funds provided by an entity that DOE chooses to administer the program. The program sunsets in 2018. Prize categories include technology advancements in hydrogen production, storage, distribution, and use; prototypes of hydrogen vehicles and products; and technologies that "transform" distribution or production. DOE is required to report to Congress annually, identifying award recipients, technologies developed, and specific actions undertaken to commercialize the technologies. More than $1 billion is authorized over a 10-year period.

Section 655 directs DOE to create the "Bright Tomorrow" lighting prizes for solid state (LED) lighting developments that achieve targeted levels of energy efficiency and other traits. Two specific categories are a solid state replacement for a 60-watt incandescent light and a replacement for the PAR Type 38 halogen light. Also, a prize is established for a "twenty-first century lamp" that achieves certain output, efficiency, and color targets. After the awards are made, DOE is required to develop guidelines for federal agency purchases of the incandescent and halogen replacements, with the goal of complete replacement within five years.

Section 656 directs DOE to establish a cost-shared Renewable Energy Innovation Manufacturing Partnership Program to make awards to support RD&D on advanced manufacturing processes, materials, and infrastructure for renewable energy technologies. Further goals are to increase domestic renewable energy production and better coordinate federal, state, and private resources through partnerships. Solar, wind, biomass, geothermal, energy storage, and fuel cell systems are eligible forms of equipment.

TITLE VII: CARBON CAPTURE AND SEQUESTRATION

Subtitle A, Carbon Capture and Sequestration Research, Development, and Demonstration

DOE's program for carbon capture and sequestration R&D is expanded and will include large-scale demonstration projects. DOE is directed to engage the National Academy of Sciences (NAS) to conduct a review of the program. DOE is directed to work with the NAS to develop interdisciplinary graduate degree programs with emphasis on geologic sequestration science. A university-based R&D grant program will be established to study carbon capture and sequestration using various types of coal. EPA is directed to assess potential impacts of carbon sequestration on public health and safety and the environment. Further, injection and sequestration activities under this subtitle are subject to the requirements of the Safe Drinking Water Act.

Subtitle B, Carbon Capture and Sequestration Assessment and Framework

Section 711 directs the Department of the Interior (DOI) to develop a methodology for an assessment of the national potential for geologic storage of carbon dioxide. Following publication of the methodology, DOI will be required to complete an assessment of national capacity for carbon dioxide storage in accordance with the methodology.

Section 712 directs DOI to develop a methodology for an assessment of the total capacity of ecosystems to sequester carbon and the ability of ecosystems to reduce emissions of carbon dioxide, methane, and nitrous oxides in ecosystems through management practices. Following publication of the methodology, DOI will be required to complete a national assessment of the quantity of carbon stored in and released from ecosystems, and the annual flux of carbon dioxide, methane, and nitrous oxides in and out of ecosystems.

Section 713 calls for DOI to maintain records, and an inventory, of the quantity of carbon dioxide stored within federal mineral leaseholds.

Section 714 directs DOI to submit a report on a recommended regulatory framework for managing geologic carbon sequestration on public lands. The report must include an assessment of options to ensure that the United States receives fair market value for the use of public land, the proposed procedures for public review and comment, procedures for protecting natural and cultural resources of the public land overlying the geologic sequestration sites, a description of the status of liability issues related to the storage of carbon dioxide in public land, identification of legal and regulatory issues for cases where the United States owns title to the mineral resources but not the overlying land, identification of issues related to carbon dioxide pipeline rights-of-way, and recommendations for additional legislation that may be required for adequate public land management and leasing to accommodate geologic sequestration of carbon dioxide and pipeline rights-of-way.

Title VIII: Improved Management of Energy Policy

Subtitle A, Management Improvements

Section 801 directs DOE to conduct a 10-year national media campaign to educate consumers to save energy and reduce oil use. Competitive bidding is required for contracting the media services. A funding authorization of $5 million per year is provided for five years. An annual report to Congress is required.

Section 802 authorizes the Federal Coordinator for the Alaska Natural Gas Transportation Projects to appoint and terminate personnel and to pay appointed and temporary personnel up to a maximum of the level III rate of the Executive Schedule. The Federal Coordinator is granted authority to establish various payment requirements and to use funds raised without further appropriation. This authority does not affect the authority of the Secretary of the Interior.

Section 803 creates a 50% matching grant program for constructing small renewable energy projects that will have an electrical generation capacity less than 15 megawatts. Eligible applicants include local governments, utilities, and Indian tribes. Such sums as necessary are authorized for the program.

Section 804 requires the Energy Information Administration (EIA) to monitor planned petroleum refinery outages and report to the Secretary of Energy when such outages are affecting the price or availability of petroleum products. The Secretary will then be required to share data with refinery operators and encourage reductions in out-of-service refinery capacity.

Section 805 requires the Administrator of the Energy Information Administration (EIA) to develop a five-year plan for enhancing the scope, quality, and timeliness of the agency's data collection efforts. In addition, it requires closer coordination by EIA with state energy officials and with the Federal Energy Regulatory Commission. The section addresses state-level data in several respects and requires the Administrator to submit to Congress within a year an assessment of state level energy data needs. EIA is directed to revisit certain data series that had been terminated due to budget constraints and to identify data gaps that may have resulted from those terminations. To implement this section, $10 million is authorized for 2008, and additional sums are authorized through 2012.

Section 806 expresses the sense of Congress that there is a national goal to use renewable energy resources from agricultural, forestry, and working lands of the nation to provide at least 25% of the nation's energy use by 2025.

Section 807 directs the Department of the Interior's U.S. Geological Survey to conduct a comprehensive assessment of geothermal energy resources in the United States and report the findings of that assessment to Congress.

Subtitle B, Prohibitions on Market Manipulation and False Information

This subtitle prohibits crude oil and petroleum product wholesalers from using any technique to manipulate the market or provide false information. The law directs the Federal Trade Commission to treat such action as an unfair or deceptive practice, subject to civil penalties of not more than $1 million per incident.

Title IX: International Energy Programs

This title authorizes assistance to promote clean and efficient energy technologies in foreign countries, and it establishes an International Clean Energy Foundation.

Subtitle A, Assistance to Promote Clean and Efficient Energy Technologies in Foreign Countries

The U.S. Agency for International Development (USAID) is directed to report to Congress on efforts to support policies for clean and efficient energy technologies. The Department of Commerce is directed to increase efforts to export such technologies and report to Congress on the results. Other U.S. agencies with export promotion responsibilities are required to increase efforts to support these technologies. Also, a multi-agency Task Force on International Cooperation for Clean and Efficient Energy Technologies is created to support the implementation of clean energy markets in key developing countries.

Section 917 creates a U.S.-Israel Energy Cooperation partnership to support research, development, and deployment (RD&D) of energy efficiency and renewable energy measures.

Subtitle B, International Clean Energy Foundation

The Foundation is established with the long-term goal of reducing GHG emissions. It is directed to use the funds authorized by this subtitle to make grants to promote projects outside of the United States that serve as models of how to reduce emissions. An annual report to Congress is required.

Subtitle C, Miscellaneous Provisions

Section 931 calls for the Secretary of State to ensure that energy security is integrated into the core mission of the Department of State. Energy advisors are required at key embassies, and the Department is required to report to Congress every two years on its energy-related activities. Section 932 adds the Secretary of Energy to the National Security Council. Section 933 calls for the President to submit to Congress a comprehensive annual report that describes a national energy security strategy for the nation.

Section 934 implements the Convention on Supplementary Compensation for Nuclear Damage that was opened for signature in 1997. The convention has since been signed by the United States and 12 other countries but has not yet entered into force. Each party to the convention will be required to establish a compensation system within its borders for nuclear damages to the public. In the United States, this obligation will be fulfilled by the existing Price-Anderson Act

(§170 of the Atomic Energy Act of 1954). The convention will also establish a second tier of damage compensation to be paid by all parties. Section 934 requires the U.S. contribution to the second tier to be paid by suppliers of nuclear equipment and services, under a formula to be developed by DOE. Supporters of the convention contend that it will help U.S. exporters of nuclear technology by establishing a predictable international liability system.

Section 935 has the stated purpose of improving national energy security by promoting anti-corruption initiatives in oil and natural gas rich countries and of improving global energy security by promoting programs such as the Extractive Industries Transparency Initiative (EITI) that aim to increase transparency and accountability into extractive resource payments. The sense of Congress is expressed that global energy security should be furthered by encouraging further participation in EITI by eligible countries and companies and by promoting the effectiveness of the EITI program by ensuring that a robust and candid review mechanism is put in place. The Secretary of State is required to report to Congress on progress made in promoting transparency in extractive industries resource payments. An authorization of $3 million is provided to support U.S. contributions to the Multi-Donor Trust Fund of EITI.

Title X: Green Jobs

This title authorizes up to $125 million in funding to establish national and state job training programs, administered by the Department of Labor, to help address job shortages that are impairing growth in green industries, such as energy efficient buildings and construction, renewable electric power, energy efficient vehicles, and biofuels development.

Title XI: Energy Transportation and Infrastructure

Subtitle A, Department of Transportation (DOT)

An Office of Climate Change and Environment is established at DOT to plan, coordinate, and implement strategies to reduce transportation-related energy use, mitigate the effects of climate change, and address the impact of climate change on transportation systems and infrastructure.

Subtitle B, Railroads

This subtitle directs DOT, in coordination with EPA, to establish and conduct a pilot grant program to assist railroad carriers in purchasing hybrid locomotives including hybrid switch locomotives, in order to demonstrate the extent to which such locomotives increase fuel economy, reduce emissions, and lower costs of operation. Also, DOT is directed to create a program of capital grants for the rehabilitation, preservation, or improvement of railroad track (including roadbed, bridges, and related track structures) of class II and class III railroads.

Subtitle C, Marine Transportation

Short sea transportation is defined as commercial waterborne transportation that originates at a port in the United States and ends at another port in the United States or at a port in Canada located in the Great Lakes Saint Lawrence Seaway System. The same definition applies for the case where origination and end points are reversed. [8] This subtitle directs DOT to establish a short sea transportation program and designate short sea transportation projects to be conducted under the program to mitigate landside congestion. Short sea shipping activities are made eligible for support from DOT's capital construction fund. A report to Congress on the short sea transportation program is required.

Subtitle D, Highways

Section 1131 increases the federal share for congestion mitigation and air quality (CMAQ) projects up to 100% of project or program cost. Under Section 1132, DOT is directed to redistribute within each state any unobligated balances of the Highway Trust Fund that are rescinded in FY2008 or FY2009. Section 1133 expresses a sense of Congress that, in constructing new roadways or rehabilitating existing facilities, state and local governments should employ policies designed to accommodate all users, including motorists, pedestrians, cyclists, transit riders, and people of all ages and abilities.

Title XII: Small Business Energy Programs

Loans, grants, and debentures are established to help small businesses develop, invest in, and purchase energy efficient buildings, fixtures, equipment, and technology. Section 1201 empowers the Small Business Administration (SBA) to make "express" loans for certain energy efficiency and renewable energy projects. Section 1202 creates a two-year pilot loan program for purchasing energy efficient technologies under Section 7(a) of the Small Business Act at half the cost that would have otherwise been required. After the pilot program terminates, GAO is required to prepare a report to Congress that describes its energy-saving impact. Section 1203 creates small business energy efficiency, sustainability, and telecommuting programs. Reports to Congress are required for each of those programs. Section 1204 raises the Small Business Investment Act (SBIA) loan ceilings for certain energy efficiency and renewable energy projects undertaken by small businesses. Section 1205 enables qualified small business investment companies to issue energy-saving debentures.

Section 1206 expands certain SBIA provisions to include investments in energy- saving small businesses. Section 1207 creates a Renewable Fuel Capital Investment (RFCI) pilot program that taps into venture capital to help small firms develop renewable energy sources and new technologies. A funding authorization of $30 million is provided for RFCI over two years. Section 1208 requires SBA to study the RFCI program and issue a report to Congress on its findings.

Title XIII: Smart Grid

Section 1301 establishes a federal policy to modernize the electric utility transmission and distribution system to maintain reliability and infrastructure protection. The term "Smart Grid" refers to a distribution system that allows for flow of information from a customer's meter in two directions: both inside the house to thermostats, appliances, and other devices, and from the house back to the utility. [9] Smart Grid is defined to include a variety of operational and energy measures — including smart meters, smart appliances, renewable energy resources, and energy efficiency resources. Section 1302 calls for DOE to report to Congress on the deployment of Smart Grid technologies and any barriers to deployment. Section 1303 directs DOE to establish a Smart Grid Advisory Committee and a Smart Grid Task Force to assist with implementation. Section 1304 directs DOE to conduct Smart Grid RD&D and to develop measurement strategies to assess energy savings and other aspects of implementation. Section 1305 directs the National Institute of Standards and Technology to establish protocols and standards to increase the flexibility of use for Smart Grid equipment and systems. Section 1306 directs DOE to create a program that reimburses 20% of qualifying Smart Grid investments. Section 1307 directs states to encourage utilities to employ Smart Grid technology and allows utilities to recover Smart Grid investments through rates. Section 1308 requires DOE to prepare a report to Congress on the effect of private wire laws on the development of combined heat and power facilities. Section 1309 directs DOE to report to Congress on the potential impacts of Smart Grid deployment on the security of electricity infrastructure and operating capability. (For additional information, see CRS Report RL34288, Smart Grid Provisions in H.R. 6, 110th Congress, by Amy Abel.)

TITLE XIV: POOL AND SPA SAFETY

Section 1401 identifies this title as the "Virginia Graeme Baker Pool and Spa Safety Act." Section 1402 finds that proper use of barriers or fencing could substantially reduce the number of childhood residential swimming and pool drownings. Section 1403 provides several definitions employed throughout this subtitle. Section 1404 sets an industry standard (ASME/ANSI A1 12.19.8) as a national performance standard for swimming pool and spa drain cover equipment. Section 1405 establishes a grant program and requires that at least 50% of the funding be used to assist states in hiring and training enforcement personnel to implement and enforce standards. The remaining funds must be used to educate pool construction and installation companies, pool owners and operators, and pool service companies, about the standard. Also, funding of $2 million per year over two years is authorized for the federal Consumer Product Safety Commission to implement the grant program. Section 1406 specifies minimum state law requirements to qualify for a grant under Section 1405. The requirement includes the enclosure of all outdoor residential pools and spas, installation of devices to prevent entrapment by pool or spa drains, and notification to pool owners about entrapment protection standards. Also, in setting minimum state law requirements, the Commission is directed to consider current or revised national standards for barrier and entrapment equipment, and to ensure that the requirements are consistent with the Commission's existing publications on pool safety guidelines. Section 1407 directs the Commission to conduct a public education program on methods to prevent drowning and entrapment in swimming pools. A funding authorization of $5 million per year is provided over five years. Section 1408 directs the Commission to submit a report to Congress that evaluates the implementation of the state grant program.

TITLE XV: REVENUE PROVISIONS

Section 1500 specifies that, unless expressed otherwise, all tax provisions in this act refer to provisions of the Internal Revenue Code of 1986.

Section 1501 extends Federal Unemployment Tax Act (FUTA) taxes for one year. FUTA imposes a 6.2% gross tax rate on the first $7,000 paid annually by covered employers to each employee. In 1976, Congress passed a temporary surtax of 0.2% of taxable wages to be added to the permanent FUTA tax rate. The temporary surtax was subsequently extended through 2007. The President's FY2008 Budget had proposed extending the FUTA surtax. The Treasury Department stated that "extending the surtax will support the continued solvency of the federal unemployment trust funds and maintain the ability of the unemployment system to adjust to any economic downturns." This section enacts the President's proposal for one year, 2008. This provision is estimated to raise $ 1.446 billion over 10 calendar years.

Under Section 1502, the geological and geophysical costs of a major integrated oil company will be amortized (deducted proportionally) over a seven-year period instead of the current five-year period. (A major integrated oil company is defined as one with an average world production of at least 500,000 barrels per day, with 2005 gross receipts exceeding $1 billion, and which has at least a 15% interest in refinery operations.)

TITLE XVI: EFFECTIVE DATE

Section 1601 specifies that this act and the amendments it makes will take effect one day after enactment.

REFERENCE

[1] For more details about the reserve, see CRS Report RS2257 1, The Strategic Energy Efficiency and Renewables Reserve in the CLEAN Energy Act of 2007 (H.R. 6, by Fred Sissine).

[2] S. 1321 was, in turn, derived from several other Senate bills. Additional details about legislation that was incorporated into H.R. 6 are available in CRS Report RL3383 1, Energy Efficiency and Renewable Energy Legislation in the 110th Congress, by Fred Sissine, Lynn J. Cunningham, and Mark Gurevitz.

[3] More details about the House and Senate bills developed up to that point are available in CRS Report RL34 135, Omnibus Energy Efficiency and Renewable Energy Legislation: A Side-by-Side Comparison of Major Provisions in House-Passed H.R. 3221 with Senate- Passed H.R. 6, by Fred Sissine.

[4] Office of Management and Budget, Statement of Administration Policy on H.R. 6, Energy Independence and Security Act of 2007, December 6, 2007; [http://www.whitehouse.gov/ omb/legislative/sap/11 0-1/hr6sap-h_2.pdf].

[5] See 42 U.S.C. 7545(f).

[6] Geopressured resources are geothermal deposits of hot water or steam found in sedimentary rocks under higher than normal pressures and that are saturated with oil and gas.

[7] These "high cost" regions are defined as places where the average cost of retail power exceeds 150% of the national average.

[8] This is the definition offered in Section 1122. Also, see the definition provided in Short Sea Shipping: Practices, Opportunities, and Challenges, by Gary A. Lombardo, [http://www.insourceaudit.com/WhitePapers/Short_Sea_Shipping.asp].

[9] The Smart Grid could allow appliances to be turned off during periods of high electrical demand and cost and give customers real-time information on constantly changing electric rates. The goal is to use advanced, information-based technologies to increase power grid efficiency, reliability, and flexibility, and reduce the rate at which additional electric utility infrastructure needs to be built.

In: Renewable Fuel Standard Issues
Editor: Daniel T. Crowe

ISBN 978-1-60692-289-7

Chapter 4

SELECTED ISSUES RELATED TO AN EXPANSION OF THE RENEWABLE FUEL STANDARD (RFS)*

Brent D. Yacobucci and Randy Schnepf

ABSTRACT

High petroleum and gasoline prices, concerns over global climate change, and the desire to promote domestic rural economies have greatly increased interest in biofuels as an alternative to petroleum in the U.S. transportation sector. Biofuels, most notably corn ethanol, have grown significantly in the past few years as a component of U.S. motor fuel supply. Ethanol, the most commonly used biofuel, is blended in nearly half of all U.S. gasoline (at the 10% level or lower in most cases). However, current biofuel supply only represents about 4% of total gasoline demand.

While recent proposals have set the goal of significantly expanding biofuel supply in the coming decades, questions remain about the ability of the U.S. biofuel industry to meet rapidly increasing demand. Current U.S. biofuel supply relies almost exclusively on ethanol produced from Midwest corn. In 2006, 20% of the U.S. corn crop was used for ethanol production. To meet some of the higher ethanol production goals would require more corn than the United States currently produces, if all of the envisioned ethanol was made from corn.

---

* Excerpted from CRS Report RL34265, dated December 3, 2007.

Due to the concerns with significant expansion in corn-based ethanol supply, interest has grown in expanding the market for biodiesel produced from soybeans and other oil crops. However, a significant increase in U.S. biofuels would likely require a movement away from food and grain crops as feedstocks. Other biofuel feedstock sources, including cellulosic biomass, are promising, but technological barriers make their future uncertain. Therefore, some proposals would require ever-larger amounts of biofuels produced from feedstocks other than corn starch, including sugarcane, oil crops, and cellulose, to promote the development of these fuels.

The Senate-passed energy bill (H.R. 6) would require the use of 36 billion gallons of renewable fuels annually in 2022, of which only 15 billion gallons could be ethanol from corn starch. The remaining 21 billion gallons would be so-called "advanced biofuels." The current RFS would only require the use of 7.5 billion gallons in 2012, increasing to an expected 8.6 billion gallons in 2022, of which only 250 million gallons of cellulosic biofuels would be required.

Issues facing the U.S. biofuels industry include potential agricultural "feedstock" supplies, and the associated market and environmental effects of a major shift in U.S. agricultural production; the energy supply needed to grow feedstocks and process them into fuel; and barriers to expanded infrastructure needed to deliver more and more biofuels to the market. A key question is whether a renewable fuel mandate is the most effective policy to promote the above goals. This report outlines some of the current supply issues facing biofuels industries, including implications for agricultural feedstocks, infrastructure concerns, energy supply for biofuel production, and fuel price uncertainties.

INTRODUCTION

High petroleum and gasoline prices, concerns over global climate change, and the desire to promote domestic rural economies have raised interest in biofuels as an alternative to petroleum in the U.S. transportation sector. Biofuels, most notably corn-based ethanol, have grown significantly in the past few years as a component of U.S. motor fuels. For example, nearly half of all U.S. gasoline contains some ethanol (mostly blended at the 10% level or lower). However, current supply represents only about 4% of annual gasoline demand on a volume basis, and only about 3% on an energy basis. In 2006, the United States consumed roughly 5 billion gallons of biofuels (mostly ethanol); this 5 billion gallons was blended into roughly 65 billion gallons of gasoline. Total annual gasoline consumption is roughly 140 billion gallons.

Recent proposals, including President Bush's goal in his 2007 State of the Union Address, aim to expand biofuel supply significantly in the coming decades. The President's goal would be to expand consumption from 5 billion gallons in 2007 to 35 billion gallons in 2017. While this proposal included not just biofuels but alternative fuels in general (including fuels from coal or natural gas), it would likely mean a significant growth in biofuels production over the next 10 years. Other legislative proposals would require significant expansion of biofuels production in the coming decades; some proposals would require 30 billion gallons of biofuels alone by 2030 or 60 billion gallons by 2050. The Senate-passed energy bill (H.R. 6) would require the use of 36 billion gallons in 2022, including 21 billion gallons of "advanced biofuels." [1]

Current U.S. biofuel supply relies almost exclusively on ethanol produced from Midwest corn (Table 1). Other fuels that play a smaller role include ethanol from Brazilian sugar, biodiesel from U.S. soybeans, and ethanol from U.S. sorghum. A significant increase in U.S. biofuels would likely require a movement away from food and grain crops. For example, U.S. ethanol production in 2006 consumed roughly 20% of the U.S. corn crop. If only corn is used, expanding ethanol production to 35 billion gallons would require more corn than the United States currently produces, which would be infeasible. Corn (and other grains) have myriad other uses, and such a shift would have drastic consequences for most agricultural markets, including grains (since corn would compete with other grains for land), livestock (since the cost of animal feed would likely increase), and land (since total harvested acreage would likely increase). In addition to agricultural effects, such an increase in corn-based ethanol would likely affect fuel costs (since biofuels tend to be more expensive than petroleum fuels), energy supply (natural gas is a key input into corn production), and the environment (since the expansion of corn-based ethanol production raises many environmental questions). These concerns are discussed below.

Due to the concerns with significant expansion in corn-based ethanol supply, interest has grown in expanding the market for biodiesel (a diesel substitute produced from vegetable and animal oils) and spurring the development of motor fuels produced from cellulosic materials (including grasses, trees, and agricultural and municipal wastes). However, all of these so-called "advanced biofuel" technologies are currently even more expensive than corn-based ethanol (with the exception of ethanol produced from Brazilian sugarcane).

Table 1. U.S. Production of Biofuels from Various Feedstocks

Fuel	Feedstock	U.S. Production in 2006
Ethanol	Corn	4.9 billion gallons
	Sorghum	less than 100 million gallons
	Cane sugar	No production (656 million gallons imported from Brazil and Caribbean countries)
	Cellulose	No production (one demonstration plant in Canada)
Biodiesel	Soybean oil	approximately 200 million gallons
	Other vegetable oils	less than 10 million gallons
	Recycled grease	less than 10 million gallons
	Cellulose	No production
Methanol	Cellulose	No production
Butanol	Cellulose, other biomass	No production

Sources: Renewable Fuels Association; National Biodiesel Board; CRS analysis.

In addition to expanding domestic production of biofuels, there is some interest in expanding imports of sugar-based ethanol from Brazil and other countries. However, ethanol from Brazil is currently subject to a $0.54 per gallon tariff that in most years is a significant barrier to direct Brazilian imports. Some Brazilian ethanol can be brought into the United States duty free if it is dehydrated (reprocessed) in Caribbean Basin Initiative (CBI) countries. [2] Up to 7% of the U.S. ethanol market could be supplied duty-free in this fashion, although historically ethanol dehydrated in CBI countries has only represented about 2% of the total U.S. market. [3]

Any fuel produced from biological materials (e.g., food crops, agricultural residues, municipal waste) is generally referred to as a "biofuel." More specifically, the term generally refers to liquid transportation fuels. The most significant biofuel in the United States is ethanol produced from U.S. corn. [4] Approximately 4.9 billion gallons of ethanol were produced in the United States in 2006, mostly from corn. Other domestic feedstocks for ethanol include grain sorghum and sweet sorghum; imported ethanol (656 million gallons in 2006) is usually produced from sugar cane in Brazil. Ethanol is generally blended into gasoline at the 10% level (E10) or lower. Ethanol can be used in purer forms such as E85 (85% ethanol and 15% gasoline) in vehicles specially designed for its use, although E85 represents less than 1% of U.S. ethanol consumption.

After ethanol, biodiesel is the next most significant biofuel in the United States, although 2006 U.S. production is estimated at only about 225 million gallons. Biodiesel is a diesel fuel substitute produced from vegetable and animal oils (mainly soybean oil in the United States), as well as recycled cooking grease. Other biofuels with the potential to play a role in the U.S. market include ethanol and diesel fuel substitutes produced from various biomass feedstocks containing cellulose. However, these "cellulosic" biofuels are currently prohibitively expensive relative to conventional ethanol and biodiesel. Other potential biofuels include other alcohols (e.g., methanol and butanol) produced from biomass.

This report outlines some of the current issues related to a significant expansion of the renewable fuel standard (RFS) established in the Energy Policy Act of 2005 (P.L. 109-58), including implications for agricultural feedstocks, infrastructure constraints, environmental concerns, energy supply issues, and fuel prices.

The Renewable Fuel Standard (RFS)

The Current RFS

Section 1501 of the Energy Policy Act of 2005 (P.L. 109-58) established a renewable fuel standard (RFS) requiring the use of renewable fuels in gasoline. Starting in 2006, the law requires the use of 4.0 billion gallons of renewable fuel in gasoline, increasing to 7.5 billion gallons in 2012. After 2012, the minimum renewable content for a given year must be no lower than the percentage of renewable fuel in gasoline in 2012. Also, starting in 2013, at least 250 million gallons of the mandate must be met using cellulosic biofuel. Although the RFS has been called an "ethanol mandate," there is no explicit requirement to use ethanol. For example, there are specific provisions for generating credits through the use of biodiesel and other renewable fuels. That said, it is expected that the vast majority of the RFS will be met using ethanol produced from corn starch.

RFS as Public Policy. An expansion in the RFS could have significant policy implications. Issues include questions of energy/petroleum security, pollutant and greenhouse gas emissions, agricultural commodity and food market effects, and infrastructure costs. Proponents and opponents of mandated biofuel use both claim that an RFS would promote or run counter to the general public interest on several different policy fronts. For example, supporters of an RFS claim it would serve several public policy interests including:

- reduced investment risk by guaranteeing demand for a projected period (such risk would otherwise keep significant investment capital on the sidelines);
- enhanced energy security via the production of liquid fuel from a renewable domestic source resulting in decreased reliance on imported fossil fuels (the U.S. currently imports over half of its petroleum, two-thirds of which is consumed by the transportation sector); and
- enhanced environmental benefits (non-toxic, biodegradable, reduced GHG emissions compared with fossil fuels, etc.).

Critics of an RFS have taken issue with many specific aspects of biofuel production and use, but a general public policy criticism of the RFS is that, by picking the "winner," policymakers may exclude or retard the development of other, potentially more preferable alternative energy sources. [5] They contend that biofuels are given a huge advantage via billions of dollars of annual subsidies which distort investment markets by redirecting venture capital and other investment dollars away from competing alternative energy sources. Instead, these critics have argued for a more "technology neutral" policy such as a carbon tax, a cap-and-trade system of carbon credits, or a floor price on imported petroleum.

The Proposed RFS Defined

This report examines the specific issues regarding the implementation of an extended RFS as contained in the Senate-passed energy bill (H.R. 6), but does not address the broader public policy issue surrounding how best to support U.S. energy policy. [6] The proposed RFS expansion would include all motor fuel, as well as heating oil (Table 2). It would expand the current RFS mandate to 13.2 billion gallons (bgal.) in 2012 (compared with the current RFS of 7.5 bgal.); 15 bgal. by 2015; and 36 bgal. in 2022. However, the corn-based ethanol share of the proposed RFS would be capped at 15 bgal. Starting in 2016, the proposed RFS would increase the RFS by an additional 3 bgal. annually, to be derived entirely from "advanced biofuels" — defined as biofuels derived from feedstocks other than corn starch. The advanced biofuel share increases to 21 bgal. by 2022. An additional requirement associated with the proposed RFS is that renewable fuels produced in facilities that commence operation after enactment must achieve at least a 20% reduction in life- cycle greenhouse gas emissions relative to gasoline.

Table 2. Proposed H.R. 6 Expansion of the Renewable Fuel Standard

Year	Current RFS (billion gallons)	Biofuel mandate for motor fuel, home heating oil, and boiler fuel (billion gallons)	Portion to be from advanced biofuel (i.e., not corn starch) (billion gallons)	Cap on corn starch-derived ethanol (billion gallons)
2006	4.0			
2007	4.7	—		—
2008	5.4	8.5		8.5
2009	6.1	10.5		10.5
2010	6.8	12.0		12.0
2011	7.4	12.6		12.6
2012	7.5	13.2		13.2
2013	7.6 (est.)[a]	13.8		13.8
2014	7.7 (est.)[a]	14.4		14.4
2015	7.8 (est.)[a]	15.0		15.0
2016	7.9 (est.)[a]	18.0	3.0	15.0
2017	8.1 (est.)[a]	21.0	6.0	15.0
2018	8.2 (est.)[a]	24.0	9.0	15.0
2019	8.3 (est.)[a]	27.0	12.0	15.0
2020	8.4 (est.)[a]	30.0	15.0	15.0
2021	8.5 (est.)[a]	33.0	18.0	15.0
2022	8.6 (est.)[a]	36.0	21.0	15.0

[a] Starting in 2013, a minimum of 250 million gallons must be derived from cellulosic biomass. After 2012, the ratio of renewable fuel to gasoline must at least equal the ratio in 2012.

The proposed RFS involves two distinct components — a corn-starch-ethanol RFS and a non-corn-starch-ethanol RFS — that are best analyzed separately because the various supply and demand factors affecting their development also are fairly distinct.

Potential Issues with an Expanded RFS

Overview of Long-Run Corn Ethanol Supply Issues

The U.S. ethanol industry has shown rapid growth in recent years, with national production increasing from 1.8 billion gallons in 2001 to 4.9 billion in 2006. This rapid growth — which is projected to continue for the foreseeable future — has important consequences for U.S. and international fuel, feed, and food markets.

Corn accounts for about 98% of the feedstocks used in ethanol production in the United States. USDA estimates that 3.2 billion bushels of corn (or 24% of the 2007 corn crop) will be used to produce ethanol during the September 2007 to August 2008 corn marketing year. [7] As of November 27, 2007, existing U.S. ethanol plant capacity was a reported 7.2 billion gallons per year, with an additional capacity of 6.2 billion gallons under construction. [8] Thus, total annual U.S. ethanol production capacity in existence or under construction as of November 27, 2007, was 13.4 billion gallons. This production capacity is well in excess of the 7.5 billion gallon supply required in 2012 by the Renewable Fuel Standard (Energy Policy Act of 2005 (P.L. 109-58)). The current pace of plant construction suggests that annual corn-forethanol use will likely require more than 4 billion bushels in 2008 and approach, or possibly exceed, 5 billion bushels by 2010. [9] In 2007, U.S. corn production was a record 13.2 billion bushels.

The ethanol-driven surge in corn demand has been associated with a sharp rise in corn prices. For example, the futures contract for March 2007 corn on the Chicago Board of Trade rose from $2.50 per bushel in September 2006 to a contract high of over $4.16 per bushel in January 2007 (a rise of 66%). Although a record U.S. corn harvest eased upward pressure on corn prices slightly during 2007, by November 2007 prices for 2008 futures contracts were again trading at over $4.00 per bushel. Both USDA and the Food and Agricultural Policy Research Institute (FAPRI) (Table 3), in their annual agricultural baseline reports, project corn prices to remain well above $3.00 per bushel through 2016 compared with an average farm price of $2.15 during the previous 10-year period (1997-2006).

This sharp rise in corn prices owes its origins largely to increasing corn demand spurred by the rapid expansion of corn-based ethanol production capacity in the United States since mid-2006. The rapid growth in ethanol capacity has been fueled by both strong energy prices and a variety of government incentives, regulations, and programs. Major federal incentives include a tax credit of $0.51 to fuel blenders for every gallon of ethanol blended with gasoline; the Renewable

Fuel Standard; and the 54¢ per gallon most-favored-nation duty on most imported ethanol. [10] A recent survey of federal and state government subsidies in support of ethanol production reported that the total annual federal support fell somewhere in the range of $5.1 to $6.8 billion per year. [11]

Market participants, economists, and biofuels skeptics have begun to question the need for continued large federal incentives in support of ethanol production, particularly when the sector would have been profitable during much of 2006 and 2007 without such subsidies. [12] Their concerns focus on the potential for widespread unintended consequences that might result from excessive federal incentives adding to the rapid expansion of ethanol production capacity and the demand for corn to feed future ethanol production. These questions extend to issues concerning the ability of the gasoline-marketing infrastructure to accommodate more ethanol in fuel, the likelihood of modifications in engine design, the environmental impacts, and other considerations.

Overview of Non-Corn-Starch-Ethanol RFS Issues

Although most references to "advanced biofuels" involve cellulosic ethanol, the "advanced biofuels" component of the proposed RFS extension may be met by essentially any non-corn-starch-derived biofuel. News reports often refer to cellulosic ethanol as "nearing a break-through" or "just around the corner" but the reality is that there is considerable uncertainty about the speed with which this technology will become commercially viable (even with substantial government support). Many scientists still suggest that commercial realization of cellulosic ethanol is 5 to 15 years down the road. [13] Although research is ongoing, presently there are no commercial-scale cellulosic biofuel plants in the United States, and there is only one demonstration-scale plant in Canada. [14] A major barrier to cellulosic fuel production is that production costs remain significantly higher than for corn ethanol or other alternative fuels. Currently, various production processes are prohibitively expensive, including physical, chemical, enzymatic, and microbial treatment and conversion of these feedstocks into motor fuel.

Unintended Policy Incentives of the "Advanced Biofuels" Mandate. The uncertainty concerning cellulosic ethanol production raises the stakes for a government mandate on its use. If cellulosic ethanol production is unable to advance rapidly enough to meet the RFS mandate for non-corn-starch ethanol, then other unexpected biofuel sources may step in and fill the void, such as:

- domestic sorghum-starch ethanol whose production may expand across the prairie states and in other regions less suitable for corn production;
- costly domestic sugar-beet ethanol or even costlier domestic biodiesel production may be undertaken to fill the mandate; or
- imports of Brazilian sugar-cane ethanol could expand.

Potential Benefits Are Vast. Ethanol and biodiesel produced from cellulosic feedstocks, such as prairie grasses and fast-growing trees or agricultural waste, have the potential to improve the energy and environmental effects of U.S. biofuels while offering significant cost savings on the production side (e.g., high-yielding, grown on marginal land, perennial rather than annual). Further, moving away from feed and food crops to dedicated energy crops could avoid some of the agricultural supply and price concerns associated with corn ethanol (as discussed later in this report).

A key potential benefit of many of cellulosic feedstocks is that many can be grown without the need for chemicals. Reducing or eliminating the need for chemical fertilizers would address one of the largest energy inputs for corn-based ethanol production. Using biomass to power a biofuel production plant could further reduce fossil fuel inputs. Improving the net energy balance of ethanol would also reduce net fuel-cycle greenhouse gas emissions.

Cellulosic Biofuel Production Uncertainties. There are substantial uncertainties regarding both the costs of production for such feedstock as well as the costs of producing biofuel from them. Perennial crops are often slow to establish and can take several years before a marketable crop is produced. Crops heavy in cellulose tend to be bulky and represent significant problems in terms of harvesting, transporting, and storing. Seasonality issues involving the operation of a biofuel plant year-round based on a 4- or 5-month harvest period of biomass suggest that bulkiness is likely to matter a great deal. In addition, most marginal lands (i.e., the low-cost biomass production zones) are located far from major urban markets, bringing the plant location choice versus ethanol transportation issue into play.

Furthermore, increases in per-acre yields would be required to make most cellulosic energy crops for fuel production economically competitive. Questions remain whether high yields can be achieved without the use of fertilizers and pesticides. Another question is whether there is sufficient feedstock supply available. USDA estimates that, by 2030, 1.3 billion tons of biomass could be available for bioenergy production (including electricity from biomass, and fuels from corn and cellulose). [15] From that, enough biofuels could be produced to replace roughly 70 billion gallons of gasoline per year (about 4.5 million barrels per day). However, this projection assumes significant increases in per-acre yields and, according to USDA, should be seen as an upper bound on what is possible. Further, new harvesting machinery would need to be developed to guarantee an economic supply of cellulosic feedstocks. [16]

In addition to the above concerns, other potential environmental drawbacks associated with cellulosic fuels must be addressed, such as the potential for soil erosion, runoff, and the spread of invasive species (many potential biofuel crops are invasive species when introduced into non-native localities). In the near term, the obvious choice of using corn stover to fuel existing corn ethanol plants has its own set of environmental trade-offs, paramount of which is the dilemma of sacrificing soil fertility gains from no- or minimum-tillage corn production.

Energy Supply Issues

Biofuels are not primary energy sources. Energy stored in biological material (through photosynthesis) must be converted into a more useful, portable fuel. This conversion requires energy. The amount and types of energy used to produce biofuels, and the feedstocks for biofuel production, are of key concern. Because of the input energy requirements, the energy and environmental benefits of corn ethanol, particularly, may be limited.

Energy Balance. A frequent argument for the use of ethanol as a motor fuel is that it reduces U.S. reliance on oil imports, making the U.S. less vulnerable to a fuel embargo of the sort that occurred in the 1970s. However, while corn ethanol use displaces petroleum, its overall effect on total energy consumption is less clear. To analyze the net energy consumption of ethanol, the entire fuel cycle must be considered. The fuel cycle consists of all inputs and processes involved in the development, delivery and final use of the fuel. For corn-based ethanol, these inputs include the energy needed to produce fertilizers, operate farm equipment, transport corn, convert corn to ethanol, and distribute the final product. Some studies find a significant positive energy balance of 1.5 or greater — in other words, the energy contained in a gallon of corn ethanol is 50% higher than the amount of energy needed to produce and distribute it. However, other studies suggest that the amount of energy needed to produce ethanol is roughly equal to the amount of energy obtained from its combustion. A review of research studies on ethanol's energy balance and greenhouse gas emissions found that most studies give corn-based ethanol a slight positive energy balance of about 1.2. [17]

If, instead, biomass was used to produce biofuels, the energy balance could be improved. It is expected that most biofuel feedstocks other than corn will require far less nitrogen fertilizer (produced from natural gas). Further, if biomass were used to provide process energy at the biofuel refinery, then the energy balance could be even greater. Some estimates are that cellulosic ethanol could have an energy balance of 8.0 or more. [18] Similarly high energy balances have been calculated for sugarcane ethanol and biodiesel.

An expanded RFS would certainly displace petroleum consumption, but the overall effect on fossil fuel consumption is questionable, especially if there is a large reliance on corn-based ethanol. The mandate in the Senate bill to require an increasing amount of "advanced biofuels" would likely result in reduced fossil fuel consumption relative to gasoline. As the share of advanced biofuels grows, this effect would increase. However, by 2022, advanced biofuels will likely represent less than 10% of gasoline energy demand, so the total amount of fossil energy displaced would be less than the expected growth in fossil energy consumption from passenger transportation over the same time period. [19]

Natural Gas Demand. As ethanol production increases, the energy needed to process the corn into ethanol, which is derived primarily from natural gas in the United States, can be expected to increase. For example, if the entire 4.9 billion gallons of ethanol produced in 2006 used natural gas as a processing fuel, it would have required an estimated 240 to 290 billion cubic feet (cu. ft.) of natural gas. [20] If the entire 2006 corn crop of 10.5 billion bushels were converted into ethanol, the energy requirements would be equivalent to approximately 1.4 to 1.7 trillion cu. ft. of natural gas. This would have represented about 6% to 8% of total U.S. natural gas consumption, which was an estimated 22.2 trillion cu. ft. in 2005. [21] The United States has been a net importer of natural gas since the early 1980s. A significant increase in its use as a processing fuel in the production of ethanol — and a feedstock for fertilizer production — would likely increase prices and imports of natural gas.

The Senate RFS proposal would likely promote an increase in corn ethanol to 15 billion gallons by 2015, requiring an increase in natural gas and/or fertilizer consumption. After 2015, the bill would promote a growing demand for fuels less dependent on natural gas.

Energy Security. [22] Despite the fact that ethanol displaces gasoline, the benefits to energy security from ethanol are not certain. As stated above, while roughly 20% of the U.S. corn crop is used for ethanol, ethanol only accounts for approximately 2% of gasoline consumption on an energy equivalent basis. [23] The import share of U.S. petroleum consumption was estimated at 54% in 2004, and is expected to grow to 70% by 2025. [24] Further, as long as ethanol remains dependent on the U.S. agricultural supplies, any threats to these supplies (such as drought), or increases in crop prices, would negatively affect the supply and/or cost of biofuels. In fact, that happened when high corn prices caused by strong export demand in 1995 contributed to an 18% decline in ethanol production between 1995 and 1996.

Further, expanding corn-based ethanol production to levels needed to significantly promote U.S. energy security is likely to be infeasible. If the entire 2007 U.S. corn crop of 13.2 billion bushels were used as ethanol feedstock, the resultant 35 billion gallons of ethanol (23.6 billion gasoline-equivalent gallons (GEG)) would represent about 16.7% of estimated national gasoline use of approximately 141 billion gallons. [25] In 2007, an estimated 86 million acres of corn were harvested (largest since 1944). Nearly 137 million acres would be needed to produce enough corn (20.5 billion bushels) and resulting ethanol (56.4 billion gallons or 37.8 billion GEG) to substitute for roughly 20% of petroleum imports. [26] Thus, barring a drastic realignment of U.S. field crop production patterns, corn-based ethanol's potential as a petroleum import substitute appears to be limited by crop area constraints, among other factors. [27]

If an expanded RFS requires a significant amount of advanced biofuels, then the specific definition of "advanced biofuel" could affect the overall energy security picture for biofuels. For example, if ethanol from sugarcane is allowed under an expanded RFS (as under the Senate bill), this could provide an incentive to increase imports of sugarcane ethanol, especially from Brazil. If not, then the expanded RFS might provide an incentive for imports of biodiesel and other renewable diesel substitutes from tropical countries.

Energy Prices. The effects of an expanded RFS on energy prices are uncertain. If wholesale biofuels prices remain higher than gasoline prices (when all economic incentives are taken into account), then mandating higher and higher levels of biofuels would likely lead to higher gasoline pump prices. However, if petroleum prices — and thus, gasoline prices — remain high, the use of some biofuels might help to mitigate high gasoline prices.

Current costs are so high for some biofuels, especially cellulosic biofuels and biodiesel from algae, that significant technological advances (or even greater increases in petroleum prices) are necessary to lower their production costs to make them competitive with gasoline. Without such cost reductions, mandating large amounts of these fuels would likely raise fuel prices. If a price were placed on greenhouse gas emissions — a policy not proposed in the Senate energy bill — then the economics could shift in favor of these fuels despite their high production costs, as they have lower fuel-cycle and life-cycle greenhouse gas emissions (see below).

Greenhouse Gas Emissions

Biofuels proponents argue that a key benefit of biofuel use is a decrease in greenhouse gas (GHG) emissions. However, some question the overall GHG benefit of biofuels, especially corn-based ethanol. There is a wide range of fuel-cycle estimates for greenhouse gas reductions from corn-based ethanol. However, most studies have found that corn-based ethanol reduces fuel-cycle GHG emissions by 10%-20% per mile relative to gasoline. [28] These estimates vary depending on several factors including the cultivation practice (e.g., minimum-tillage versus normal tillage) used to grow the corn and the fuel used to process the corn into ethanol (e.g., natural gas versus coal). These same studies find that biofuels produced from sugar cane or cellulosic biomass could reduce fuel-cycle GHG emissions by as much as 90% per mile relative to gasoline.

However, fuel-cycle analyses generally do not take changes in land use into account. For example, if a previously uncultivated piece of land is tilled to plant biofuel crops, some of the carbon stored in the field could be released. In that case, the overall GHG benefit of biofuels could be compromised. One study estimates that taking land use into account (a life-cycle analysis, as opposed to a fuel-cycle analysis), the GHG reduction from corn ethanol is less than 3% per mile relative to gasoline, [29] while cellulosic biofuels have a life-cycle reduction of 50%. [30]

The Senate bill requires that to qualify under the expanded RFS, biofuels produced at facilities commencing operations after the date of enactment must have a 20% life-cycle emissions reduction. However, it is expected that this provision may not be relevant to a large share of conventional ethanol since much of the capacity to meet the 15 billion gallon cap is currently existing or will come from expansions of existing plants. If enactment of the Senate bill is delayed (or if plants currently under construction are grandfathered under the bill) the GHG reductions from conventional ethanol may be insignificant. However, the more "advanced biofuel" required under the bill, the greater the likely benefit in terms of GHG reductions.

Agricultural Issues

A continued expansion of corn-based ethanol production could have significant consequences for traditional U.S. agricultural crop production and rural economies. Supporters of an expanded RFS claim that increased biofuels production and use would have enormous agricultural and rural economic benefits by increasing farm and rural incomes and generating substantial rural employment opportunities. [31] However, large-scale shifts in agricultural production activities will likely also have important regional economic consequences that have yet to be fully explored or understood. As corn prices rise, so too does the incentive to expand corn production either by expanding onto more marginal soil environments or by altering the traditional corn-soybean rotation that dominates Corn Belt agriculture. This shift could displace other field crops, primarily soybeans, and other agricultural activities. Further, corn production is among the most energy-intensive of the major field crops. An expansion of corn area would likely have important and unwanted environmental consequences due to the increases in fertilizer and chemical use and soil erosion. The National Corn Growers Association estimates that U.S. corn-based ethanol production could expand to between 12.8 and 17.8 billion gallons by 2015 without significantly affecting agricultural markets. [32] However, as noted below, other evidence suggests effects are already being felt in the current expansion in corn production.

Food versus Fuel. Many critics of federal biofuels subsidies and the RFS argue that a sustained rise in grain prices driven by ethanol feedstock demand likely will lead to higher U.S. and world food prices with potentially harmful effects on consumer budgets and nutrition. [33] As evidence they cite USDA's estimate that the U.S. Consumer Price Index (CPI) for all food is forecast to increase 3.5% to 4.5% in 2007 compared with an increase of 2.4% in 2006 and, an average annual rate of 2.5% over the past ten years (1997-2006). [34] However, in analyzing this critique it is important to distinguish between prices of farm-level crops and retail-level food products because most "food" prices are largely determined by costs and profits after the commodities leave the farm. [35] Basic economics suggests that the price of a particular retail food item varies with a change in the price of an underlying ingredient in direct relation to the relative importance (in value terms) of that ingredient. For example, if the value of wheat in a $1.00 loaf of bread is about 10¢, then a 20% rise in the price of wheat translates into a 2¢ rise in a loaf of bread.

As a result of corn's relatively small value-share in most retail food product prices, it is unlikely that the ethanol-driven corn price surge is a major factor in current food price inflation estimates. [36] Furthermore, economists generally agree that most retail food price increases are not due to ethanol-driven demand increases, but rather are the result of two major factors — a sharp increase in energy prices which ripples through all phases of marketing and processing channels, and the strong increase in demand for agricultural products in the international marketplace from China and India (a product of their large populations and rapid economic growth). [37]

Table 3. U.S. Farm Prices for Major Agricultural Commodities

		Farm Market Prices				USDA Program Prices[g]	
		Average	Actual	Projections			Target
Commodity	Unit	1997-2006	2006/07	2007/08[d]	2012/13[e]	Loan rate	Price
Wheat[a]	$/bu	3.24	4.26	6.20	4.29	2.75	3.92
Corn[a]	$/bu	2.15	3.04	3.50	3.25	1.95	2.63
Sorghum[a]	$/bu	2.04	3.29	3.30	3.02[f]	1.95	2.57
Barley[a]	$/bu	2.38	2.85	3.85	3.11[f]	1.85	2.44
Oats[a]	$/bu	1.54	1.87	2.40	1.90[f]	1.33	1.44
Rice[a]	$/cwt	7.17	9.74	11.00	9.64	6.50	10.50
Soybeans[a]	$/bu	5.72	6.43	9.00	7.72	5.00	5.80
Soybean oil[b]	¢/lb	21.4	31.0	39.5	36.8	—	—
Soybean meal[b]	$/st	187.7	205.4	250.0	202.0	—	—
Cotton, Upland	¢/lb	50.3	46.5	53.9[e]	59.9	52.0	72.4
Choice Steers[c]	$/cwt	73.5	85.4	91.6	86.4	—	—
Barrows/Gilts[c]	$/cwt	42.2	47.3	47.0	54.3	—	—
Broilers[c]	¢/lb	37.9	64.4	76.1	77.2	—	—
Eggs[c]	¢/doz	63.7	71.8	109.1	85.4[f]	—	—
Milk[c]	$/cwt	13.91	12.90	19.00	15.7	—	—

a Season average farm price from USDA, National Agricultural Statistical Service, Agricultural Prices. — = no loan rate.

b USDA, Agr. Marketing Service (AMS), Decatur, IL, cash price, simple average crude for soybean oil, and simple average 48% protein for soybean meal.

c Calendar year data for the first year, e.g., 2000/2001 = 2000; USDA, AMS: choice steers — Nebraska, direct 1100-1300 lbs.; barrows/gilts — national base, live equivalent 51%-52% lean; broilers — wholesale, 12-city average; eggs — Grade A, New York, volume buyers; and milk — simple average of prices received by farmers for all milk.

d Unless otherwise indicated: mid-point of price projection range from USDA, World Agricultural Supply and Demand Estimates (WASDE) Nov. 9, 2007;.

e Unless otherwise indicated: FAPRI, Baseline Update for U.S. Agricultural Markets, August 2007.

f FAPRI, U.S. Baseline Briefing Book, February 2007, FAPRI-UMC Report #02-07.

g For more information on U.S. commodity programs see CRS Report RL33271, Farm Commodity Programs: Direct Payments, Counter-Cyclical Payments, and Marketing Loans.

Feed Markets. Most corn grown in the United States is used for animal feed. From 1995 through 2005, domestic feed use accounted for 58% of U.S. corn use. As corn-based ethanol production increases, so do total corn demand and corn prices. As a result, prolonged higher corn prices likely will have significant consequences for traditional feed markets and the livestock industries — hog, cattle, dairy, and poultry — that depend on those feed markets. Corn traditionally has represented about 57% of feed concentrates and processed feedstuffs fed to animals in the United States. [38] Persistently high feed costs will tighten profit margins and likely squeeze out marginal livestock producers. Because economies of scale tend to favor larger producers, persistently tighter profit margins suggest a potential for increased concentration in the livestock sector. The National Cattlemen's Beef Association (NCBA) has been one of the foremost critics of an expanded RFS. Instead, the NCBA argues for a phase out of current ethanol subsidies and a more market-based approach to renewable fuels policy. [39]

The price of corn also is linked to the price of other grains, including those destined for food markets, through competition in the feed marketplace and in the producer's planting choices for limited acreage. The price runup in the U.S. corn market has already spilled over into price increases in the markets for soybeans and soybean oil. Supply distortions also are likely to develop in protein-meal markets related to expanded production of the ethanol processing by-product distiller's dried grains with solubles (DDGS), which averages about 30% protein content and can substitute in certain feed and meal markets. [40]

While DDGS use would substitute for some of the lost feed value of corn used in ethanol processing, about 66% of the original weight of corn is consumed in producing ethanol and is no longer available for feed. Furthermore, not all livestock species are well adapted to dramatically increased consumption of DDGS in their rations — dairy cattle appear to be best suited to expanding DDGS's share in feed rations; poultry and pork are much less able to adapt. Also, DDGS must be dried before it can be transported long distances, adding to feed costs. There may be some potential for large-scale livestock producers to relocate near new feed sources, but such relocation likely would have important regional economic effects.

Domestic Food Prices. Although corn primarily is used as a livestock feed or for ethanol production, corn also is used widely as an ingredient (albeit minor) in many processed foods, e.g., soft drinks, snack foods, and baked goods. Since corn prices are a relatively small share of the price of most retail food products, their price impact is concomitantly small. Higher corn prices have their largest impact on meat prices. The feed-price effect will first translate into higher prices for poultry and hogs, which are less able to use alternate feedstuffs. Dairy and beef cattle are more versatile in their ability to shift to alternate feed sources, but eventually a sustained rise in corn prices will push their feed costs upward as well. A recent economic study estimated that a 30% increase in the price of corn, and associated increases in the prices of wheat and soybeans, would increase egg prices by 8.1%, poultry prices by 5.1%, pork prices by 4.5%, beef prices by 4.1%, and milk prices by 2.7%. [41] The effect on all food consumed was a 1.1% increase (0.9% on at-home food and 1.3% on away-from-home food consumption). Thus, the price impact of higher corn prices is small but important for most livestock products, and probably much smaller for most other retail food products.

The overall impact to consumers from higher food prices depends on the proportion of income that is spent on food. Since food costs represent a relatively small share of consumer spending for most U.S. households (about 10%), food price increases (from whatever source) are absorbed relatively easily in the short run. However, low-income consumers spend a much greater proportion of their income on food than do high-income consumers. Their larger share combined with less flexibility to adjust expenditures in other budget areas means that any increase in food prices potentially could cause hardship. [42] In addition, higher commodity prices combined with shrinking inventories mean that local school districts and the U.S. government will be forced to pay higher market prices for food for school lunch programs. And the automatic food price escalators built into the food stamp program mean rising expenditures as well. [43]

International Food Prices. Due to trade linkages, the increase in U.S. corn prices has become a concern for international markets, as well. High commodity prices ripple through international markets where impacts vary widely based on grain import dependence and the ability to respond to higher commodity prices. Import- dependent developing country markets are put at greater food security risk due to the higher cost of imported commodities. In particular, lower-income households in many foreign markets where food imports are an important share of national consumption and where food expenses represent a larger portion of the household budget may be affected by higher food prices. [44] In China, where corn is an important food source, the government recently has put a halt to its planned ethanol plant expansion due to the threat it poses to the country's food security. Similarly, humanitarian groups have expressed concern for the potential difficulties that higher grain prices imply for developing countries that are net food importers. [45]

Exports. The United States is the world's leading exporter of corn. In the past decade (1997 to 2006), the United States has exported about 20% of its corn production, accounting for nearly 66% of world corn trade. [46] Increased use of corn for ethanol production could diminish U.S. capacity for exports. In 2006, the volume of corn used for ethanol equaled exports, with a 20% share of total use. By the 2009/10 marketing year (September-August), ethanol's share of U.S. corn production is expected to reach nearly 36%, while the export share falls to 13%. [47] FAPRI projections clearly suggest that higher corn prices will result in lost export sales. It is unclear what type of market adjustments will occur in global feed markets, since several different grains and feedstuffs are relatively close substitutes. Price-sensitive corn importers may quickly switch to alternative, cheaper sources of feed, depending on the availability of supplies and the adaptability of animal rations. In contrast, less price-sensitive corn importers, such as Japan and Taiwan, may choose to pay a higher price in an attempt to bid the corn away from ethanol plants. There could be significant economic effects to U.S. grain companies and to the U.S. agricultural sector if ethanol-induced higher corn prices led to a sustained reshaping of international grain trade.

Economic Impact. Several studies claim that increased biofuels production and use would have enormous agricultural and rural economic benefits by increasing farm and rural incomes and generating substantial rural employment opportunities. [48] One estimate suggested that the economic benefit from the ethanol industry to the U.S. economy for 2005 was $17.7 billion of GDP; the creation of over 150,000 jobs; $5.7 billion in spinoff economic activity, and more than $3.5 billion in government tax revenues. [49] However, a recent critical review of the standard input-output methodology used to generate such economic impact estimates suggests that the income and job growth attributable to biofuels production has been grossly overstated (perhaps by as much as a factor of 4 or 5). [50] Yet, while the magnitude may be called into question, there appears to be no doubt about the potential positive value of biofuels production to rural economies. First, in addition to temporary construction work to build a new plant, several dozen permanent jobs also accompany a biofuel plant (the eventual job number depends on the size of the plant's operating capacity). Second, the new demand boosts the local prices received by farmers for corn and sorghum. Third, important secondary economic activity is associated with the operation of an ethanol plant. Fourth, given the high level of federal and state subsidies for the biofuels industry, any locality that is home to a biofuels plant can expect substantial net transfers of government funds into the area's economy.

The policy question of interest is not whether there are positive gains from growth in the ethanol industry, but whether the growth and its economic implications are sufficient to merit large government subsidies. A growing number of critics argue that the answer is no. [51] Other suggest that, at the very least, the issue deserves more study before expanding current government support levels. [52]

Infrastructure and Distribution Issues

In addition to the above concerns about raw material supply for biofuel production (both feedstock and energy), there are issues involving biofuel distribution and infrastructure. Expanding ethanol production likely will strain the existing supply infrastructure. Further, expansion of ethanol use beyond certain levels will require investment in entirely new infrastructure that would be necessary to handle a higher and higher percentage of ethanol in gasoline. If biomass-based diesel substitutes are produced in much larger quantities, some of these infrastructure issues may be mitigated.

Distribution Issues. Ethanol-blended gasoline tends to separate in pipelines. Further, ethanol is corrosive and may damage existing pipelines. Therefore, unlike petroleum products, ethanol and ethanol blended gasoline cannot be shipped by pipeline in the United States. Another issue with pipeline transportation is that corn ethanol must be moved from rural areas in the Midwest to more populated areas, which are often located along the coasts. This shipment is in the opposite direction of existing pipeline transportation, which moves gasoline from refiners along the coast to other coastal cities and into the interior of the country. While some studies have concluded that shipping ethanol or ethanol-blended gasoline via pipeline could be feasible, no major U.S. pipeline has made the investments to allow such shipments. [53]

Thus, the current distribution system for ethanol is dependent on rail cars, tanker trucks, and barges. These deliver ethanol to fuel terminals where it is blended with gasoline before shipment via tanker truck to gasoline retailers. However, these transport modes lead to higher prices than pipeline transport, and the supply of current shipping options (especially rail cars) is limited. For example, according to industry estimates, the number of ethanol carloads has tripled between 2001 and 2006, and the number is expected to increase by another 30% in 2007. [54] A significant increase in corn-based ethanol production would further strain this tight transport situation.

Because of these distribution issues, some pipeline operators are seeking ways to make their systems compatible with ethanol or ethanol-blended gasoline. These modifications could include coating the interior of pipelines with epoxy or some other, corrosion-resistant material. Another potential strategy could be to replace all susceptible pipeline components with newer, hardier components. However, even if such modifications are technically possible, they likely will be expensive, and could further increase ethanol transportation costs.

If non-corn biofuels play a larger role, as required in the Senate bill, some of the supply infrastructure concerns may be alleviated. Cellulosic biofuels potentially can be produced from a variety of feedstocks, and may not be as dependent on a single crop from one region of the country. For example, municipal solid waste is ubiquitous across the United States, and could serve as a ready feedstock for biofuel production if the technology were developed to convert it to fuel economically. Further, increased imports of biofuels from other countries could allow for greater use of biofuels, especially along the coasts.

Higher-Level Ethanol Blends. One key benefit of gasoline-ethanol blends up to 10% ethanol is that they are compatible with existing vehicles and infrastructure (e.g., fuel tanks, retail pumps, etc.). All automakers that produce cars and light trucks for the U.S. market warranty their vehicles to run on gasoline with up to 10% ethanol (E10). This 10% currently is an upper bound to the amount of ethanol that can be introduced into the gasoline pool. If most or all gasoline in the country contained 10% ethanol, this would allow only for roughly 15 billion gallons, far less than the amount of biofuel mandated in the Senate bill.

As a major producer of ethanol for its domestic market, Brazil has a mandate that all of its gasoline contain 20-25% ethanol. For the United States to move to E20 (20% ethanol, 80% gasoline), it may be that few (if any) modifications would need to be made to existing vehicles and infrastructure. Vehicle testing, however, would be necessary to determine whether new vehicle parts would be required, or if existing vehicles are compatible with E20. Similar testing would be necessary for terminal tanks, tanker trucks, retail tanks, pumps, etc. In addition, EPA would need to certify that the fuel will not lead to increased air quality problems.

There is also interest in expanding the use of E85 (85% ethanol, 15% gasoline). Current E85 consumption represents only about 1% of ethanol consumption in the United States. A key reason for the relatively low consumption of E85 is that relatively few vehicles operate on E85. The National Ethanol Vehicle Coalition estimates that there are approximately six million E85-capable vehicles on U.S. roads, [55] as compared to approximately 230 million gasoline- and diesel-fueled vehicles. [56] Most E85-capable vehicles are "flexible fuel vehicles" or FFVs. An FFV can operate on any mixture of gasoline and between 0% and 85% ethanol. However, owners of a large majority of the FFVs on U.S. roads choose to fuel them exclusively with gasoline, largely due to higher per-mile fuel cost [57] and lower availability of E85.

E85 capacity is expanding rapidly, with the number of E85 stations roughly doubling between February 2006 and February 2007. But those stations still represent less than 1% of U.S. gasoline retailers. Further expansion will require significant investments, especially at the retail level. If a new E85 pump and underground tank are necessary, they can cost as much as $100,000 to $200,000 to install. [58] However, if existing equipment can be used with little modification, the cost could be less than $10,000.

Vehicle Infrastructure Issues. As was stated above, if a large portion of any increased RFS is met using ethanol, then the United States likely does not have the vehicles to consume the fuel. The 10% limit on ethanol in gasoline for conventional vehicles poses a significant barrier to expanding ethanol consumption beyond 15 billion gallons per year. [59] To allow more ethanol use, vehicles will need to be certified and warrantied for higher-level ethanol blends, or the number of ethanol FFVs will need to increase.

CONCLUSION

There is continuing interest in expanding the U.S. biofuel industry as a strategy for promoting energy security and environmental goals. However, there are limits to the amount of biofuels that can be produced from current feedstocks and questions about the net energy and environmental benefits they would provide. Further, rapid expansion of biofuel production may have many unintended and undesirable consequences for agricultural commodity costs, fossil energy use, and environmental degradation. As policies are implemented to promote ever-increasing use of biofuels, the goal of replacing petroleum use with agricultural products competes with these other potential consequences. Further, alternative strategies for energy conservation and alternative energy production are widely seen as warranting consideration.

REFERENCE

[1] The term "advanced biofuel" comes from current legislation in the 1 10th Congress, including the Senate energy bill (H.R. 6). In many cases, the definition of "advanced biofuel" includes mature technologies and fuels that are currently produced in large amounts. For example, the H.R. 6 definition of "advanced biofuel" includes ethanol from sugarcane, despite the fact that Brazilian sugar growers have been producing fuel ethanol for decades. H.R. 6 defines "advanced biofuels" as biofuels other than ethanol derived from corn starch. Possible fuels include biodiesel from oil seeds, ethanol from

sugarcane, and ethanol from cellulosic materials (including non-starch parts of the corn plant, such as the stalk).

[2] For more information on CBI imports, see CRS Report RS2 1930, Ethanol Imports and the Caribbean Basin Initiative, by Brent D. Yacobucci.

[3] In 2006 ethanol prices rose sharply, and direct imports from Brazil rose sharply, despite the tariff.

[4] For more information on ethanol, see CRS Report RL33290, Fuel Ethanol: Background and Public Policy Issues, by Brent D. Yacobucci.

[5] 5 For example, see Bruce A. Babcock, "High Crop Prices, Ethanol Mandates, and the Public Good: Do They Coexist?" Iowa Ag Review, Vol. 13, No. 2, Spring 2007; and Robert Hahn and Caroline Cecot, "The Benefits and Costs of Ethanol," Working Paper 07-17, AEI- Brookings Joint Center for Regulatory Studies, November 2007.

[6] The same RFS extension is also proposed in the Domenici Amendment (S.Amdt. 3614) to the Senate Agriculture Committee's approved farm bill (S. 2302).

[7] USDA, WAOB, World Agricultural Supply and Demand Estimates (WASDE) Report, Nov. 9, 2007, Washington; available at [http://www.usda.gov/oce/].

[8] See Renewable Fuels Association, Industry Statistics, at [http://www.ethanolrfa.org/ industry/statistics/].

[9] FAPRI, Baseline Update for U.S. Agricultural Markets, FAPRI-MU report #28-07,August 2007.

[10] For more information on incentives (both tax and non-tax) for ethanol, see CRS Report RL3 3572, Biofuels Incentives: A Summary of Federal Programs, by Brent D. Yacobucci.

[11] Koplow, Doug. Biofuels — At What Cost? Government Support for Ethanol and Biodiesel in the United States, Global Subsidies Initiative of the International Institute for Sustainable Development, Geneva, Switzerland, October 2006; available at [http://www.globalsubsidies.org].

[12] Chris Hurt, Wally Tyner, and Otto Doering, Department of Agricultural Economics, Purdue University, Economics of Ethanol, December 2006, West Lafayette, IN.

[13] For example, the Department of Energy's goal is to make cellulosic biofuels cost- competitive with corn ethanol by 2012. Other groups are less optimistic.

[14] However, on February 28, 2007, DOE announced availability of $385 million in grant funding for six commercial-scale cellulosic ethanol plants in six states. If operational, combined capacity of these six plants would be

130 million gallons per year. DOE, DOE Selects Six Cellulosic Ethanol Plants for Up to $385 Million in Federal Funding, February 28, 2007, Washington, D.C.

[15] Oak Ridge National Laboratory for DOE and USDA, Biomass as a Feedstock for a Bioenergy and Bioproducts Industry: The Technical Feasibility of a Billion-Ton Annual Supply, April 2005, Oak Ridge, TN.

[16] For example, the study assumes roughly 400 million tons of biomass from agricultural residues. To economically supply those residues to biofuel producers, farm equipment manufacturers likely would need to develop one-pass harvesters that could collect and separate crops and crop residues at the same time.

[17] Alexander E. Farrell, Richard J. Plevin, Brian T. Turner, Andrew D. Jones, Michael O'Hare, and Daniel M. Kammen, "Ethanol Can Contribute to Energy and Environmental Goals," Science, Jan. 27, 2006, pp. 506-508.

[18] David Andress, Ethanol Energy Balances. November 2002.

[19] For example, EIA projects that motor gasoline consumption will increase 22% between 2007 and 2011. EIA, Annual Energy Outlook. Table 11.

[20] CRS calculations based on energy usage rates of 49,733 Btu/gal of ethanol from Shapouri (2004), roughly 60,000 Btu/gal from Farrell (2006). Hosein Shapouri and Andrew McAloon, USDA, Office of the Chief Economist, The 2001 Net Energy Balance of Corn- Ethanol, 2004, Washington; Farrell, op. cit.

[21] U.S. Department of Energy (DOE), Energy Information Administration (EIA), Annual Energy Outlook 2006 with Projections to 2030, Table 1- Total Energy Supply and Disposition Summary, Washington; at [http://www.eia.doe.gov/oiaf/aeo/index.html].

[22] A key question in evaluating the energy security benefits or costs of an expanded RFS is "what is the definition of energy security." For many policymakers, "energy security" and "energy independence" (i.e., producing all energy within our borders) are synonymous. For others, "energy security" means guaranteeing that we have reliable supplies of energy regardless of their origin. For this section, the former definition is used.

[23] By volume, ethanol accounted for approximately 3.6% of gasoline consumption in the United States in 2006, but a gallon of ethanol yields only 67% of the energy of a gallon of gasoline.

[24] DOE, EIA, Annual Energy Outlook 2004 with Projections to 2025, Washington.

[25] Based on USDA's Jan. 12, 2007, World Agricultural Supply and Demand Estimates (WASDE) Report, and using comparable conversion rates.

[26] This represents roughly half of gasoline's share of imported petroleum. However, petroleum imports are primarily unrefined crude oil, which is then refined into a variety of products. CRS calculations assume corn yields of 150 bushels per acre and an ethanol yield of 2.75 gal/bu.

[27] Two recent articles by economists at Iowa State University examine the potential for obtaining a 10 million acre expansion in corn planting: Bruce Babcock and D. A. Hennessy, "Getting More Corn Acres From the Corn Belt"; and Chad E. Hart, "Feeding the Ethanol Boom: Where Will the Corn Come From?" Iowa Ag Review, Vol. 12, No. 4, Fall 2006.

[28] EPA, Greenhouse Gas Impacts of Expanded Renewable and Alternative Fuels Use. April 2007; Farrell, et al.

[29] Mark A. Delucchi, Draft Report: Life cycle Analyses of Biofuels. 2006.

[30] While a 50% life-cycle reduction is still significant, it is far less than the 90% reduction suggested by fuel-cycle analyses.

[31] For example, see John M. Urbanchuk (Director, LECG LLC), Contribution of the Ethanol Industry to the Economy of the United States, white paper prepared for National Corn Growers Assoc., February 21, 2006.

[32] National Corn Growers Association, How Much Ethanol Can Come From Corn?, November 9, 2006, Washington D.C.

[33] For a discussion, see the National Corn Growers Association's online "Food versus Fuel Debate," at [http://www.ncga.com/news/OurView/pdf/2006/FoodANDFuel.pdf].

[34] ERS, USDA, Briefing Room "Food CPI, Prices, and Expenditures," at [http://www.ers.usda.gov/Briefing/CPIFoodAndExpenditures/consumerpriceindex.htm].

[35] Helen H. Jensen and Bruce A. Babcock, "Do Biofuels Mean Inexpensive Food is a Thing of the Past?" Iowa Ag Review, Spring 2007, Vol. 13, No. 2, pp. 1-3.

[36] For examples, see Food & Water Watch, "Retail Realities: Corn Prices Do Not Drive Grocery Inflation," Sept. 2007; and John M. Urbanchuk (Director, LECG LLC), "The Relative Impact of Corn and Energy Prices in the Grocery Aisle," white paper prepared for National Corn Growers Assoc., June, 14, 2007.

[37] For examples, see Jacque Diouf, Director General of the U.N. Food and Agriculture Organization, "Why Are Food Prices Rising?" in Financial Times Online, Nov. 26, 2007; [http://media.ft.com/cms/s/2/f5bd920c-975b-1 1dc-9e08-0000779fd2ac.html?from=textlink]. See also, Keith Collins, Chief Economist, USDA, Testimony before the House Committee on Agriculture, October 18, 2007.

[38] USDA, ERS, Feed Situation and Outlook Yearbook, FDS-2003, Apr. 2003, Washington.
[39] "NCBA on Renewable Fuel Policy," NCBA Issue Backgrounder-2007; available at [http://www.beefusa.org/uDocs/NCBAonRenewableFuelPolicy-2007.pdf].
[40] For a discussion of potential feed market effects due to growing ethanol production, see Bob Kohlmeyer, "The Other Side of Ethanol's Bonanza," Ag Perspectives (World Perspectives, Inc.), Dec. 14, 2004; and R. Wisner and P. Baumel, "Ethanol, Exports, and Livestock: Will There be Enough Corn to Supply Future Needs?," Feedstuffs, no. 30, vol. 76, July 26, 2004.
[41] Simla Tokgoz and others, "Emerging Biofuels: Outlook of Effects on U.S. Grain, Oilseed, and Livestock Markets," Staff Report 07-SR 101, Center for Agricultural Research and Development (CARD), Iowa State University, May 2007.
[42] Helen H. Jensen and Bruce A. Babcock, "Do Biofuels Mean Inexpensive Food is a Thing of the Past?" Iowa Ag Review, Spring 2007, Vol. 13, No. 2, pp. 1-3.
[43] Ibid.
[44] Shahla Shapouri and Stacey Rosen, "Energy Price Implications for Food Security in Developing Countries," Food Security Assessment, 2006, GFA-18, Economic Research Service, USDA.
[45] International Monetary Fund, World Economic Outlook: Globalization and Ineqauality. October 2007. Washington.
[46] USDA, Production, Supply and Distribution Online (PSD database) available at [http://www.fas.usda.gov/psdonline/psdHome.aspx].
[47] FAPRI, Baseline Update for U.S. Agricultural Markets, August 2007.
[48] For example, see John M. Urbanchuk (Director, LECG LLC), Contribution of the Ethanol Industry to the Economy of the United States, white paper prepared for National Corn Growers Assoc., February 21, 2006; see also Urbanchuk, Contribution of the Biofuels Industry To the Economy of Iowa, white paper prepared for the Iowa Renewable Fuels Association, Feburary 2007.
[49] Urbanchuk (2006).
[50] David Swenson, "Input-Outrageous: The Economic Impacts of Modern Biofuels Production," Dept. of Econ, Iowa State University (ISU), June 2006. Similar results are found in: David Swenson, "Understanding Biofuels Economic Impact Claims," Dept. of Econ, ISU, April 2007; Lisa Eathington and Dave Swenson, "Dude, Where's My Corn? Constraints on the Location of Ethanol Production in the Corn Belt," Dept. of Econ, ISU,

paper presented at 46th Annual Meeting of the Southern Regional Science Assoc., Charleston, SC, March 29-31, 2007; Swenson and Eathington, "Determining the Regional Economic Values of Ethanol Production in Iowa Considering Different Levels of Local Investment," Dept. of Econ, ISU, July 2006.

[51] Examples include Robert Hahn and Caroline Cecot, "The Benefits and Costs of Ethanol," Working Paper 07-17, AEI-Brookings Joint Center for Regulatory Studies, November 2007; Richard Doornbosch and Ronald Steenblik, "Biofuels: is the Cure Worse Than the Disease?", paper presented at an OECD Round Table on Sustainable Development, Paris, September 11-12, 2007; and Doug Koplow, Biofuels at What Cost? Government Support for Ethanol and Biodiesel in the United States: 2007 Update, report prepared for the Global Studies Initiative of the International Institute for Sustainable Development, Geneva, Switzerland, October 2007.

[52] For example, see Bruce A. Babcock, "High Crop Prices, Ethanol Mandates, and the Public Good: Do They Coexist?" Iowa Ag Review, Vol. 13, No. 2, Spring 2007.

[53] Some small, proprietary ethanol pipelines do exist. American Petroleum Institute, Shipping Ethanol Through Pipelines. Available at [http://www.api.org/aboutoilgas/sectors /pipeline/upload/pipelineethanolshipment-2.doc]

[54] Ilan Brat and Daniel Machalaba, "Can Ethanol Get a Ticket to Ride?," The *Wall Street Journal*, Feb. 1, 2007, p. B1.

[55] National Ethanol Vehicle Coalition, Frequently Asked Questions, accessed February 3, 2006, at [http://www.e85fuel.com/e85 101/faq.php].

[56] Federal Highway Administration, Highway Statistics 2003, November 2004, Washington.

[57] Ethanol has a lower energy content than gasoline per gallon. Therefore, FFVs tend to have lower fuel economy when operating on E85. For the use of E85 to be economical, the pump price for E85 must be low enough to make up for the decreased fuel economy relative to gasoline. Generally, to have equivalent per-mile costs, E85 must cost 20% to 30% less per gallon at the pump than gasoline.

[58] David Sedgwick, Automotive News, January 29, 2007. p. 112.

[59] Note that 15 billion gallons is the corn starch ethanol limit for the expanded RFS in the Senate-passed energy bill.

In: Renewable Fuel Standard Issues
Editor: Daniel T. Crowe

ISBN 978-1-60692-289-7

Chapter 5

Robert J. Meyers (Office of Air and Radiation, U.S. Environmental Protection Agency) Before the Committee on Environment and Public Works Subcommittee on Clean Air and Nuclear Safety United States Senate July 10, 2008

Mr. Chairman and members of the Subcommittee, I appreciate the opportunity to come before you today to testify on implementation of the renewable fuel provisions of the Energy Independence and Security Act of 2007 (EISA). The Act's aggressive new renewable fuel standards (RFS) will further our nation's goals of achieving energy security and reducing greenhouse gases by building on the successful RFS program established by the Energy Policy Act of 2005 (EPACT 2005).

Renewable fuels are a key element of a national strategy for addressing our energy security and the challenge of global climate change. The national renewable fuel standards, in combination with the vehicle fuel economy standards in EISA, will reduce emissions of greenhouse gases in the transportation sector and improve our energy security. The changes brought about by EISA are expected to prevent the release of billions of metric tons of greenhouse gases emissions into the atmosphere over the next several decades.

The Environmental Protection Agency is responsible for implementing the RFS program, and we are proud of our success to date in working with stakeholders in industry, states and the environmental community to build an effective program for increasing the volumes of renewable fuel used by the transportation sector. In April 2007 we announced final regulations for implementing the RFS Program under EPACT 2005. The Agency worked very closely with both our federal partners and stakeholders to develop broad support for the program. This program was officially launched in September 2007. We believe our success is grounded on our close collaboration with stakeholders on the design and implementation of the program.

Since EISA was signed into law on December 19, 2007, the Agency has been working diligently to develop regulations to implement the new RFS program established by that legislation, commonly called RFS2. Our first and most pressing task was to issue a new renewable volume standard for 2008. The RFS program established by EPACT 2005 required 5.4 billion gallons of renewable fuel in 2008. The EISA legislation increased the standard to 9 billion gallons in 2008, with annual increases in mandated volumes resulting in 36 billion gallons being required in 2022. We published a notice implementing the 2008 volume requirement in the Federal Register on February 14 of this year.

While the RFS program established under EPACT 2005 provides a solid foundation for the new regulations, RFS2 includes new elements which add complexity to the program. As a result, the new EISA provisions require careful evaluation and considerable new analysis.

In this new undertaking, the Agency is following much of the same approach we used in developing the first RFS program. This includes obtaining critical input from our stakeholders throughout the rulemaking process. Since EISA was enacted less than seven months ago, the Agency has met with more than thirty different stakeholders, including renewable fuel producers, technology companies, petroleum refiners and importers, agricultural associations, environmental groups, gasoline and petroleum marketers, pipeline owners and fuel terminal operators. Agency technical staff have participated in numerous conferences and workshops, which have allowed us to reach a broad range of technical, programmatic and policy issue experts. We also continue to meet and collaborate regularly with the Departments of Energy and Agriculture. Through these meetings, EPA has sought input on the key RFS2 program design elements as highlighted in this testimony.

While EPA will draw from its experience in developing the original RFS regulations, it is important to understand that EISA made a significant number of changes to the RFS program. First, as mentioned previously, RFS2 increases the total renewable fuel volumes mandated to 36 billion gallons a year by 2022. This is nearly a five fold increase over the 7.5 billion gallons a year mandated under EPACT 2005 for 2012, and constitutes a 10-year extension of the schedule provided for in that legislation. EPA believes that the implications of this substantial increase are not trivial. Development of infrastructure capable of delivering, storing and blending these volumes in new markets and expanding existing market capabilities will be needed. In addition, the market's absorption of increased volumes of ethanol will ultimately require new “outlets” beyond E10 blends (i.e., gasoline containing 10% ethanol by volume). A rule of thumb estimate is that E10 blends, if used nationwide, would utilize approximately 15 billion gallons of ethanol. Accommodating approximately an additional 20 billion gallons of ethanol- blended fuel is expected to require an expansion of the number of flexible-fuel E85 vehicles and their utilization of E85 and/or other actions. New emerging renewable fuel production technologies may hold potential to make gasoline and diesel-like fuels from renewable sources. The Agency will continue to monitor and evaluate the development of such technologies as we implement the RFS program over the coming years.

Second, beyond the significant increase in the volume mandate, EISA extended the RFS program to include both non-road gasoline and diesel fuel volumes. Under the regulations implementing EPACT 2005, RFS volume requirements were applied only to producers and importers of on-road gasoline. The extension of this program to both non- road gasoline and diesel fuel volumes, along with the potential for opt-in by participants of the home hearing oil and jet fuel markets is a significant change that may affect new parties, including a number of small businesses that have not been regulated under this program in the past.

Third, EISA has established new categories of renewable fuel. EPACT 2005 established standards for two categories of renewable fuels: one standard for the total volume of renewable fuel; and a second standard for cellulosic ethanol requiring 250 million gallons beginning in 2013. RFS2 increased the number of renewable fuel categories and standards to a total of four, including total renewable fuel and three new categories within that with unique volume requirements: advanced biofuels, biomass- based diesel and cellulosic biofuels. Industry will be required to demonstrate compliance with the four separate fuel standards. This will likely require the obligated parties, producers and importers, to forge new business relationships and contracts that are necessary to guarantee their compliance with the new standards. Establishing the necessary systems to track and verify the production and distribution of these fuels and demonstrate compliance with four separate standards will also require sufficient lead time to design and implement these new tracking systems. As in the current program under EPACT 2005, in the near term, some parties may not be able to comply by blending the renewable fuels, and thus may need to purchase or trade credits for the appropriate number and category of fuels to satisfy their volume obligations. It will be very important to conduct effective outreach with these parties to support a smooth implementation. In addition, certain requirements in RFS2 pertain only to renewable fuel production facilities that commence construction after the legislation was enacted. EPA will need to carefully consider how this new provision should be interpreted.

As part of its restructuring of the renewable fuel mandate, EISA increased the cellulosic biofuel mandate from 250 million to 1.0 billion gallons by 2013, with additional yearly increases to 16 billion gallons in 2022. EISA also provided a new definition of this fuel: cellulosic biofuel must be derived from renewable biomass, which includes requirements that place various limitations on the types of land from which the feedstocks are taken, and cellulosic biofuel must also have lifecycle greenhouse gas emissions that are at least 60 percent less than the baseline lifecycle greenhouse gas emissions for petroleum based fuel (RFS2 established the baseline year as 2005).

Implementing these requirements will entail additional work by EPA as it develops its upcoming regulation. For example, the Act authorizes EPA in certain circumstances to adjust the cellulosic biofuel standard to a level lower than that specified in the law. However it requires in this circumstance that the Agency also make credits available for compliance purposes and provides instructions on how to establish a specific price for these credits. The Agency will therefore need to address several critical issues, such as the quantity of credits to be generated, to whom they will be available, the extent to which they can be traded, and the life of the credit.

RFS2 also established for the first time minimum volume standards for biomass based diesel fuel. These standards begin in 2009 at a half billion gallons and ramp up to one billion gallons per year in 2012 and thereafter. To qualify as biomass based diesel, the renewable fuel portion of the biomass based diesel blend must result in greenhouse gas emissions that are at least 50 percent lower than the baseline GHG emissions for petroleum based diesel fuel (RFS2 established the baseline year as 2005) and cannot be co-processed with a petroleum feedstock.

Fourth, EISA requires the Agency to apply lifecycle greenhouse gas (GHG) performance threshold standards to each category of renewable fuel. Congress provided a specific definition of lifecycle analysis that requires EPA to consider all stages of fuel and feedstock production and distribution, from feedstock generation or extraction through the distribution and use of the finished fuel to the ultimate consumer. The Act also specifies that EPA take into account both direct emissions and significant indirect emissions such as emissions from land use changes.

EPA is currently developing a methodology that meets the EISA requirements. This effort builds on a substantial amount of work the Agency has done in this area, beginning with our analysis of the lifecycle GHG impact of the renewable fuel volumes required by the RFS1 program. EPA has expanded the methodology to include secondary agricultural sector impacts and land use changes. The Agency is continuing to further refine and improve our analyses as we prepare to implement the statute's lifecycle GHG performance thresholds.

Given the importance of lifecycle analysis to the success of the RFS2 program and the complexity of this work, the Agency has been working closely with stakeholders. Through multiple meetings with a broad range of groups--including the Departments of Energy and Agriculture, academics and lifecycle experts, environmental organizations, renewable fuel producers, and refiners--we have shared our approach and sought input on the key assumptions and modeling tools necessary to conduct a complete lifecycle analysis that meets the EISA criteria. These discussions have been extremely valuable to the Agency and we plan to maintain this high level of stakeholder engagement throughout the rule development process.

Fifth, RFS2 added a number of other new provisions, including changing the definition of renewable fuel feedstocks in a fundamental manner. The new law limits the crops and crop residues used to produce renewable fuel to those grown on land cleared or cultivated at any time prior to enactment of EISA, that is either actively managed or fallow, and non-forested. EISA also requires that forest-related slash and tree thinnings used for renewable fuel production pursuant to the Act be harvested from non-federal forest lands. Developing appropriate and enforceable regulations addressing these provisions requires extensive dialogue with USDA, USTR, the agricultural community and renewable fuel producers to better understand current practices and changes in practices that can be developed, implemented and enforced. The Agency has started these discussions and plans to continue this dialogue throughout the regulatory process.

Finally, in support of the rulemaking, we are assessing the many impacts of the EISA renewable fuel program. Assessments are underway to understand the impacts on emissions and air quality (greenhouse gases, ozone, particulate matter and toxics), water impacts (including water quality and consumption), agricultural sector impacts (including direct and indirect land use change), energy security, and economic impacts (such as cost of fuels and feedstocks). Detailed information will be needed for the draft regulatory impact analysis (RIA), which we intend to release with the proposed rules. These analyses will provide important information to the public and Congress on the many anticipated impacts of the new legislation.

With respect to other implementation issues, as you are aware, Texas Governor Rick Perry sent a letter to EPA Administrator Johnson on April 25 requesting a partial waiver of the 2008 RFS volume obligations required by EISA. Governor Perry requested the volume requirement be reduced by 50 percent, from 9 billion gallons in 2008 to 4.5 billion gallons, citing an "unnecessarily negative impact on Texas' otherwise strong economy while driving up global food prices". EPA issued a federal register notice on May 22, requesting public comment on this request. The comment period closed on June 23. We received over 15,000 comments, with over 150 substantive comments from a wide range of stakeholders including individual companies and associations representing renewable fuel producers, farmers, cattle, beef and poultry industries, the food and grain industries and many others. We have been evaluating these comments and other pertinent information and conducting the analysis necessary to support a decision by the Administrator. In compliance with statutory requirements, EPA is also consulting extensively with our colleagues at the Departments of Agriculture and Energy.

EPA has also been closely monitoring the aftermath of the mid-west floods to determine to what extent this natural disaster may impact the renewable fuel program. We have had multiple discussions with the USDA, DOE, renewable fuel producers, oil companies, petroleum marketers and state authorities. We are evaluating both impacts on feedstock (e.g. corn, soybeans, etc.) availability for use in ethanol production, as well impacts on fuel production and distribution systems. The extent of these impacts has not yet been fully determined. We will continue to coordinate and collaborate with DOE and USDA closely on this matter.

In closing, the Agency is moving forward with the development of regulations implementing the RFS2 provisions and is utilizing the successful approach we employed in developing the regulations for the original RFS program. We look forward to working closely with members of Congress and our many other stakeholders during this process.

Thank you, Mr. Chairman, and members of the Subcommittee for this opportunity. This concludes my prepared statement. I would be pleased to answer any questions that you may have.

In: Renewable Fuel Standard Issues
Editor: Daniel T. Crowe

ISBN 978-1-60692-289-7

Chapter 6

STATEMENT OF STEVEN G. CHALK (RENEWABLE ENERGY, U.S. DEPARTMENT OF ENERGY) BEFORE THE SUBCOMMITTEE ON CLEAN AIR AND NUCLEAR SAFETY COMMITTEE ON ENVIRONMENT AND PUBLIC WORKS UNITED STATES SENATE JULY 10, 2008

Mr. Chairman, Ranking Member Voinovich, Members of the Subcommittee, thank you for the opportunity to appear before you today to discuss the Renewable Fuel Standard (RFS) included in the Energy Independence and Security Act of 2007 (EISA), particularly life-cycle analysis of alternative fuel usage and the Department of Energy's (DOE) research and development of the next generation of biofuels. All of us recognize the national and economic security importance of reducing our dependence on oil. The implementation of the RFS is one way that Congress and the Administration have recognized and responded to the urgency of expanding use of non-petroleum fuels to improve energy security, reduce costs, and reduce greenhouse gas emissions.

EISA's Renewable Fuel Standard and Life-Cycle Analysis

In the 2007 State of the Union Address, President Bush called on Congress to significantly increase the use of advanced biofuels as part of the Twenty in Ten Initiative. Congress passed and the President signed EISA into law, requiring that U.S. transportation fuels contain at least nine billion gallons of renewable fuels in 2008, growing to 36 billion gallons in 2022. Of the quantity required in 2022, at least 21 billion gallons must be advanced biofuels (non-corn ethanol), and of that 21 billion, 16 billion gallons must be cellulosic biofuels; to meet the RFS, ethanol from corn is capped at 15 billion gallons. DOE and other federal agencies are working to develop diverse, nonfood feedstocks that require little water or fertilizer, and to foster sustainable agricultural and forestry practices. Our efforts will help spur the resources, technologies, and systems at the rate and scale needed to enable this mandate to be met.

The Department believes that the RFS is critical in scaling up the production and use of biofuels in the U.S. and deploying next generation biofuels. Creating a predictable policy environment for investors is critical to ensuring growth in all parts of the biofuels supply chain, from feedstocks, to biorefineries, to infrastructure, including pipelines. In both the short and long term, relaxing the RFS would likely undercut investments in new capacity as well as in research, development, and demonstration of cellulosic ethanol and other advanced biofuels. Generally speaking, the RFS included in the EISA can act to displace petroleum used for transportation purposes and to reduce greenhouse gas emissions; relaxing that mandate can hinder progress toward these efforts.

DOE previously projected that the transportation fuel industry had the ability to meet the nine billion gallon requirement for renewable fuels this year, from domestic ethanol and biomass-based diesel production, ethanol imports, and credits accrued by refiners for 2007 blending levels. However, the effects of recent flooding in the Midwest have not been fully analyzed, and we cannot predict future catastrophic weather events that may have an impact. Section 21 1(o)(7) of the Clean Air Act, as amended by EISA, gives the Environmental Protection Agency (EPA) Administrator the authority to waive, in whole or in part, the total volume of renewable fuel mandated by the RFS if there is inadequate domestic supply to meet the mandate, or if "implementation of the requirement would severely harm the economy of a State, a region, or the United States." The EPA Administrator is required to consult with the Department of Agriculture (USDA) and DOE if he is to consider a petition to waive or reduce the

RFS in a given year. The consultative process with DOE has begun because of a petition submitted by Texas Governor Rick Perry to reduce the RFS from 9.0 to 4.5 billion gallons in 2008.

Over the longer term, to ensure the continued availability of resources to meet RFS volume requirements in an environmentally sustainable manner, we are implementing the sustainability provisions of EISA.

- EPA, in consultation with DOE and USDA has already initiated the development of a methodology to assess the life-cycle greenhouse gas emissions of biofuels production and use to ensure conventional and advanced biofuels meet minimum greenhouse reduction requirements.
- EISA Section 204 requires EPA, in consultation with DOE and USDA to assess and report to Congress on the environmental and resource conservation impacts of the RFS every three years. EPA has initiated this work.
- EISA Section 203 requires DOE, in consultation with USDA and EPA, to contract with the National Academy of Sciences (NAS) to assess the impact of the renewable fuel standard on the industries involved in the production of feed grains, livestock, food, forest products, and energy. DOE and NAS have conducted preliminary discussions to scope this study.

The Department is focused on robust empirical validation of all environmental impacts of bioenergy across the production life cycle, from feedstocks to vehicles. In FY 2008, DOE is funding approximately $1 million worth of research across the national labs and universities to collect and analyze data to assess the indirect impacts of biofuels production. This work is to validate DOE's existing life cycle analysis models as well as inform the development of new analytical tools.

In addition to these analytical efforts, DOE is engaged in field trials to grow and harvest dedicated energy crops into biofuels, to address the lack of data available at a large scale, so that life cycle analysis tools can be adequately validated. A better understanding of carbon and water cycling in environments where dedicated energy crops are produced will increase our ability to model the environmental impacts of large-scale bioenergy production nationally.

The Biomass R&D Board, an interagency coordination group established by the Biomass R&D Act of 2000 and co-chaired by DOE and USDA, is also addressing sustainability issues. The Board recently established a Sustainability Working Group, chaired by DOE, USDA, and EPA. This group is focused on collecting and analyzing existing criteria, benchmarks, and indicators to ensure sustainable production of biofuels. Such indicators may include metrics for water quality and quantity, GHG emissions, and air quality, as well as displacement of fossil-based energy for transportation fuels.

DOE ADVANCED BIOFUELS RESEARCH, DEVELOPMENT, AND DEMONSTRATION

The Department actively supports biofuels production, from the most basic science research activities to efforts toward the integration of advanced biofuels into the national fuel supply. To help meet our long-term energy needs, the Department's biomass research and development (R&D) activities are designed to make biofuels from non-food feedstocks cost competitive by 2012.

The biomass feedstocks of today include grains (corn, sorghum, wheat), as well as oilseeds from plants (such as soybeans). Our goal is to allow the feedstocks of tomorrow to come from a variety of sources such as wastes and residues, and fast-growing energy crops. These future feedstocks may consist of agricultural residues like stalks, stems, and other crop wastes, as well as forest resources such as wood waste, forest thinnings, and small-diameter trees. Examples of energy crops include switchgrass, miscanthus, and hybrid poplar trees, in addition to algae and non-edible oilseeds like jatropha. Sorted municipal solid waste may also play a role.

Cellulosic ethanol is expected to improve upon the positive energy balance of today's corn ethanol by delivering four to six times as much energy as needed for production. [1] Additionally, DOE research has shown that cellulosic feedstocks can reduce life-cycle greenhouse gas emissions by 86 percent compared to gasoline. [2]

As a result of research, development, and demonstration efforts to date, I am pleased to report that we have already made significant progress toward the Presidential goal of making cellulosic ethanol cost competitive, which involves technology improvements to the production process that will result in a modeled cost of $1.33 (2007 dollars) per gallon of cellulosic ethanol by 2012. [3] Since 2001, the Federal Government has helped reduce the modeled cost of cellulosic production by 60 percent, from $6.00 to $2.40 per gallon.

The Department plans to invest up to $585 million over five years (FY 2007 – FY 2011), subject to appropriations, in cost-shared, integrated commercial- and demonstration-scale biorefineries that could produce up to 130 million gallons of ethanol from cellulosic biomass when fully operational. These biorefineries will demonstrate a wide range of advanced biochemical and thermochemical conversion technologies and use a wide array of cellulosic feedstocks.

On November 6, 2007, Range Fuels, Inc., became the first of the commercial-scale companies selected by DOE last February, as a part of the EPACT 2005 integrated biorefineries solicitation, to break ground on a commercial cellulosic ethanol plant, one of the first in the Nation. The plant is located near the town of Soperton, Georgia, and will draw on gasification technology to convert wood and wood waste from Georgia's pine forests and mills into 20 million gallons of ethanol per year during its first phase of operation. Construction of the first phase is expected to be completed next year.

In addition, the Department's Office of Science has recently established three major new DOE Bioenergy Research Centers—led by the University of Wisconsin-Madison, Oak Ridge National Laboratory, and Lawrence Berkeley National Laboratory, respectively— which are bringing together top scientists and researchers in an effort to accelerate the transformational breakthroughs in basic science needed to make next-generation cellulosic biofuels cost-effective. The Department plans to invest over $400 million, subject to appropriations, in the outyears.

Conclusion

The United States is now recognized as a world leader in committing to renewable fuels as a key component of its energy security strategy. Many developed and developing countries now also have renewable fuel standards and targets, including the European Union, Australia, India, and China. [4] As a recent example, on June 26, 2008, Canada, a major oil-producing nation, [5] implemented a renewable fuels standard requiring all gasoline to contain 5% biofuels by 2010 and diesel and home heating oil to contain 2% biodiesel by 2012.

We believe that the expanded RFS creates a predictable investment climate that will enable the substantial participation of the private sector, whose commitment is essential to maintaining U.S. world leadership in the development of next-generation, non-food biofuel feedstocks, and achieving the critical EISA goal of making the use of renewable fuels a standard component of the transportation fuel market.

Mr. Chairman, thank you again for holding this important hearing and for the opportunity to address EISA's renewable fuel requirements and the Department of

Energy's work in advanced biofuels. This concludes my prepared statement, and I would be happy to answer any questions the Committee Members may have.

Reference

[1] Source: Wang et al, "Life-cycle energy and greenhouse gas emission impacts of different corn ethanol plant types," Environmental Research Letters, May 2007.

[2] Ibid.

[3] Biomass Multi-Year Program Plan, Office of the Biomass Program, Energy Efficiency and Renewable Energy, DOE, March 2008, http://www1.eere.energy.gov/biomass/pdfs/biomass_program_mypp.pdf.

[4] Rajagopal and Zilberman, "Review of Environmental, Economic and Policy Aspects of Biofuels," The World Bank, September 2007.

[5] "Country Energy Profiles," EIA, http://tonto.eia.doe.gov/country/index.cfm.

In: Renewable Fuel Standard Issues
Editor: Daniel T. Crowe

ISBN 978-1-60692-289-7

Chapter 7

STATEMENT OF JOHN PIERCE (TECHNOLOGY DUPONT APPLIED BIOSCIENCES E.I. DUPONT DE NEMOURS AND COMPANY, INC) REGARDING THE RENEWABLE FUELS STANDARD BEFORE THE SUBCOMMITTEE ON CLEAN AIR AND NUCLEAR SAFETY COMMITTEE ON ENVIRONMENT AND PUBLIC WORKS U.S. SENATE JULY 10, 2008

Good morning Chairman Carper, ranking member Voinovich and members of the committee. My name is John Pierce, and I am the Vice President for Technology for DuPont Applied Biosciences, which includes our biofuels and biomaterials businesses. I am pleased to be here today to discuss the Renewable Fuels Standard. In my testimony I will provide our views of the future of biofuels, and the role of the Renewable Fuels Standard in that future.

DuPont brings a broad perspective to bear on biofuels issues, as we span the biofuels value chain. Our seed business Pioneer Hi-Bred is the world's largest seller of seed corn and the second largest seller of soybean seed to farmers. Our seeds have enabled steadily rising yields, or production per acre, for over 80 years. We sell over 170 corn varieties specifically for ethanol production, varieties that produce high fermentable starch, yielding more gallons of ethanol per acre.

With our Partner BP we are developing biobutanol, a higher alcohol fuel produced by fermenting biomass. Biobutanol is more like gasoline than is ethanol, with higher energy density and the ability to be distributed via the existing gasoline infrastructure, including pipelines. Biobutanol also improves ethanol-gasoline blends as a co-blending component. It reduces the volatility of the blend, allowing butanol-ethanol-gasoline blends to be used in the summer season where air quality concerns currently limit the use of ethanol-gasoline blends. And with our partner Genencor we will next year pilot a cellulosic ethanol technology based on corn stover, with commercial quantities produced in 2012. This technology will quickly provide expanded ethanol production from the existing agricultural and biorefinery infrastructure employing a non-food feedstock. We also have biomaterials and food ingredients businesses. Across our entire business portfolio we are experiencing the impacts of rising input prices, from energy to steel to agricultural commodities, and understand the strains they impose. That is why we are producing materials that help to reduce the supply-demand imbalances that contribute to these price rises.

The existing suite of biofuels policies, of which the RFS is a significant component, has been very successful in standing up a US ethanol industry that is making a meaningful contribution to US energy security and the environmental footprint of transportation. We have to remember that we started down the road of alternative transportation fuels because of the variety of security, environmental and economic ramifications of our dependence on petroleum. Those challenges have only grown more acute. And we are making solid progress. Today, ethanol production in the US offsets over 7 billion gallons of petroleum demand. A recent analysis from Iowa State University estimates that the presence of ethanol in the US fuel pool is lowering gasoline prices on the order of 25 to 40 cents per gallon. That is a significant savings for American consumers and businesses. Because energy costs are also a major element of manufacturing costs this is also helping to restrain inflationary pressures.

The Role of Biofuels

We believe that biofuels can serve an expanded role in fueling transportation in the US and elsewhere while contributing to reducing the carbon intensity of transportation. We also believe this can be done without increasing the environmental footprint of the agricultural enterprise. Steadily increasing agricultural productivity, which we are helping to achieve, and the use of non-food feedstocks, such as cellulose, to produce biofuels are important to this future.

Ensuring a viable biofuels market will be critical to ensuring the substantial private sector investments that are required to bring these second generation biofuels technologies to market in the next several years.

Why are we bullish on biofuels? First, because we see agricultural productivity as an engine than can provide abundant food, feed, fuel and materials globally. And second because we see the promise of next generation biofuels technologies to expand upon the solid foundation we have built on grain based ethanol. DuPont and BP will be producing biobutanol at pilot scale in 2010 and commercial scale in 2012. We have already performed fleet testing and are on track to bring to market a biofuel that is completely compatible with the existing petroleum infrastructure, has high energy density and thus good fuel mileage, and improves ethanol-gasoline blends. Next year our joint venture with Genencor will start up a pilot plant producing ethanol from corn stover, which is the cob, stalk and leaves of the corn plant, with commercial production in 2012. This fuel will have a carbon improvement over gasoline on the order of 80 to 90%.

Why corn stover? Because it capitalizes on the existing infrastructure to provide rapid expansion in ethanol production with no potential competition for food and feed uses of corn. The same equipment that goes into the field to harvest corn grain for ethanol will harvest an appropriate amount of stover, leaving behind enough for soil conditioning and erosion control. The stover will be transported to an existing biorefinery where it will be fed to a parallel processing and fermentation unit integrated with the existing facility. The result will be a 25-30% increase in ethanol production from the existing acreage. The stover from corn fields that are currently producing for food and feed uses will be able to produce additional biofuels volumes, further expanding the ability of agriculture to produce food, feed and fuel. While other cellulosic feedstocks have promise, and we are working on them as well, corn stover offers the most rapid deployment.

THE ROLE OF AGRICULTURAL PRODUCTIVITY

As for agricultural productivity, the US has a long track record of continually expanding production from the existing agricultural acreage, and producing a wide variety of products for food, feed, fuel, and industrial uses. At the turn of the 20th century, 25% of all energy used in the US came from burning wood, and in 1915, some 90 million acres of US cropland were used to grow feed for horses and mules – our transportation at that time. A major, new innovation of the last century was to learn to use fossil fuel based sources for our transportation, energy, and material needs. While hugely transformational for the world's economy, limitations of our dependency on fossil fuels are increasingly apparent. Concurrently with the huge transition to this fossil based economy, agricultural productivity also increased by leaps and bounds. When our Pioneer subsidiary began operations in 1926, corn yields were about 27 bushels per acre and petroleum was relatively cheap – you could buy 3.5 pounds of petroleum for the cost of one pound of corn. Today, corn yields in the US average about 150 bushels per acre. Corn, at $7/bushel, is 3.5 times cheaper than petroleum, instead of being 3.5 times more expensive as it was in 1926 – a remarkable testament to agricultural productivity.

Agricultural yield and productivity has steadily increased in the US, with notable gains in a few other areas of the world, allowing significantly expanded production from the same acreage (see exhibit 1). In the last 25 years improved corn yields from existing acres in the US have resulted in corn production that would have required an additional 150 million planted acres had yields not steadily improved. In essence, better yield has created 150 million "virtual acres", about the amount of planted land in the US today. In the last ten years global soybean production has increased 56% and corn production has increased 32%, while the total acres of land used for such production has increased only 6% (see exhibit 2). Such is the power of agricultural yields. This has come from steadily improving plant varieties that produce more, require less inputs and are less susceptible to insects, disease and weather variations. For example, in 1983, 1988 and 1994 the US experienced droughts that reduced average corn yield 25-30%. In the 2003 drought the yield declined only about 7%. So we are seeing higher upside and lower downside in agricultural production. Further, we expect within the next ten years to see a further 40% increase in this rate of annual gain. In 1985 average corn yield in the US was about 100 bushels per acre. In 1995 that value was 130 bushels per acre, and in 2005 it was 150 bushels per acre. We think 2020 will see average yields of 200 bushels per acre (see exhibit 3). This enhanced productivity from existing acreage with more efficient use of inputs such as water, fertilizer and crop protection chemicals also contributes to more sustainable agriculture.

However, much of the rest of the world lags US productivity by large margins, even after significant gains in some regions. In the last ten years Brazil has increased its corn yields by 50% and South Africa almost 70%, yet they still have yields less than half of those in the US, as do South and East Asia (see exhibit 4). This means that there are dramatic opportunities to expand global agricultural production from existing acreage by bringing modern farming practices to other parts of the world (see exhibit 5). This is why we believe that agriculture can continue to provide for our food, feed, fiber and fuel needs.

THE RFS

Now let me turn my attention to some of the specifics of the Renewable Fuels Standard. As the RFS was developed we encouraged Congress to emphasize policies that were feedstock, technology and fuel type neutral, and to focus on desired fuel attributes such as energy density, low carbon content and infrastructure compatibility. The RFS provisions enacted last year made significant steps in this direction, and are helping to motivate the right kinds of market transitions. In fact, there are multiple technology developers intending to produce cellulosic ethanol in pilot or demonstration quantities from a range of feedstocks over the next 24 months. The economics and carbon performance of grain ethanol continues to improve as well, as does agricultural productivity and sustainability in the US. These trends suggest that while the RFS targets are aggressive, as they should be, they are not out of reach. The integrated strategy of the U.S. is exemplified by the RFS and the related investment strategies of US DOE and USDA, which have supported a variety of alternative fuels technologies. As a result, numerous companies and institutions are now involved in biofuels work looking at a variety of different technologies, approaches, and feedstocks, and we can confidently look to US-derived technology to make the seminal contributions to renewable fuels.

We would like to see biofuels policies continue to evolve to be more performance based, to provide the right market signals for the production and use of fuels with the most beneficial attributes. For example, a low carbon fuel standard would create incentives for a fuel blender to purchase biofuels with the lowest life cycle carbon content. That would create greater market value for such low carbon fuels, and thus market incentives for fuel producers to develop the lowest carbon biofuels. Single value threshold carbon standards, such as a minimum percentage improvement over gasoline in life-cycle carbon content, run the risk of blunting that market signal. If biofuels producers don't see higher market value for fuels with better carbon performance than a bright line standard they will be less likely to develop such fuels. A low carbon fuel standard or other "sliding scale" approach better incents the market to produce the best fuels possible. However, we recognize that having an efficient and functional low carbon fuel standard will require streamlined and standardized life cycle assessment tools that are not yet available, which I will return to in a minute.

We also think it is in the national interest to help further accelerate the development of cellulosic biofuels. There are a number of things Congress can do in this regard. You can help to ease the financing risk of new biorefineries, such as through loan guarantees and accelerated deprecation. There is also a significant need to develop the know how and infrastructure for growing, harvesting, transporting, storing and processing cellulosic feedstocks, as this remains one of the least developed areas for this emerging technology. The Farm Bill made progress in these areas. Continued federal encouragement to growers and further R&D attention to cellulosic feedstocks would be very beneficial.

That being said, we think the current policy framework provides a sound basis and is helping to develop a robust biofuels industry. We look forward to continuing to work with Congress to improve this framework over time while avoiding sudden changes that would disrupt our path to broader use of biofuels. However, there are some potential storm clouds on the horizon.

BIOFUELS SUSTAINABILITY

One of the benefits of, and appropriate goals for, biofuels is enhanced sustainability, including a lower carbon footprint, than the incumbent hydrocarbon fuels. Given the nature of the biofuels value chain it is important that we consider the carbon ramifications of the various elements of that value chain, from the production of feedstock through their harvest and transportation through the biorefinery process. This is typically done through life cycle modeling, which is an area under rapid development and where common methodologies have not yet emerged. We will have to get to standardized and streamlined life cycle tools before we can fully utilize the market power of something like a low carbon fuels standard. The US currently consumes almost 150 billion gallons of gasoline a year, a staggering amount. As biofuels become an increasingly prominent part of that fuel pool, it will mean many, many batches of biofuels from many, many producers. Current life-cycle tools are quite complex, and can look more like a mini-PhD thesis than the kind of real-time tool that will be necessary for evaluating and certifying these batches of biofuels. The tool will also have to allow for ready differentiation amongst fuels on things like carbon content so fuel producers can capture value for fuels with better attributes.

Of course, the implementation of a coherent market based carbon reduction program, such as cap and trade, for fossil-based carbon emissions provides perhaps the most straightforward way to accomplish the goals of reducing fossil

CO_2 emissions in that it would incent all users of fossil-based fuels – including biofuels producers – to find ways to lower their fossil CO_2 footprint.

Congress has tasked EPA with making life cycle analysis considerations at a broad policy level as it implements the "RFS II", including the consideration of what are termed indirect land use effects. These are land use changes that are not directly attributed to the production of biofuels feedstocks but that in some manner can be inferred to be indirectly attributed.

What is well known is that converting certain kinds of ecosystems to agricultural or other use, such as tropical forests and peat bogs, can release significant amounts of stored carbon and reduce future carbon storage potential, as well as reducing habitat and having other deleterious effects. We have, for example, seen such effects in parts of Southeast Asia where forests have been felled for logging as well as for palm oil monoculture. What is much less understood is the potential for expanded agricultural production to more generically result in such land use changes, which are second, third or even fourth order effects. There are multiple factors that influence land use – human population growth, rising standards of living, traditional subsistence farming practices, global demand for timber and minerals, etc. The role of agricultural production in motivating such land use changes is an area of much speculation, including speculations from some recently publicized analyses and media stories, but very little real knowledge. While we think that understanding the negative aspects of inappropriate disturbance of these most sensitive ecosystems provides a cautionary tale of how to regulate land use and how not to produce timber, biofuels feedstocks, and other agricultural commodities, we do not see such outcomes as either desired or inevitable in our future.

It appears that EPA intends to apply potential indirect effects of biofuels feedstocks production generically to biofuels. In large measure this is due to the significant difficulty of relating a particular feedstocks production to land use changes that might occur thousands of miles away. This approach could result in some sort of generic carbon "penalty" being applied to biofuels. We are concerned that this generic carbon penalty, by discounting the actual carbon performance of a given fuel batch, could serve to mute the market signal that properly constructed policy would send to encourage the production and consumption of the lowest life-cycle carbon fuels. We think that considerations of directly attributable land use changes should be included in the life cycle considerations of particular fuels, and that the potential for indirect land use changes are best controlled through other policies that more directly address and prevent deleterious changes.

We would also caution that biofuels feedstocks are only one market for agricultural products. Farmers do not grow just for the food, feed, materials, or fuel markets. They grow-and have always grown- crops that serve all of these markets. If we begin to set standards for agricultural practices for biofuels feedstocks that differ from those for agriculture generally we run the risk of balkanizing agricultural production, and in the process creating disincentives for production for certain end uses. It would be like establishing different mail delivery standards for every tenth house on a mail route. Continually improving the sustainability of agriculture is important, and we are seeing steady improvements in agricultural practices, hardier plants, and lower inputs such as water and fertilizer. We should continue to advance the sustainability of agriculture as a whole, and not fall prey to multiple and potentially conflicting sustainability standards for crops going into different end uses. The end uses may be different, but the crops and the growing are the same.

Another area where caution is warranted is the current attribution by some of rising global commodity prices, particularly food commodity prices, to ethanol driven corn demand, and the resulting suggestion that the RFS be stalled or decreased. Many others have noted the variety of factors that have caused the prices of a wide variety of commodities, from steel and copper to cement, energy and, yes, grains to rise. Prices for foods derived from wheat, corn, and rice have all risen over a similar period, though these crops have dramatically different food, feed, and fuel uses. The primary drivers for these price increases can be found in higher demand resulting from population growth, higher per capita income and rising standards of living, and overall higher fossil fuel energy costs. I would simply observe that in 2008 ethanol production in the US is anticipated to consume about 19% of US corn production, which is about 10% of global corn production. Corn into animal feed, including exports (17% of production), will consume 70% of US production (taking into account the ethanol co-product DDG that goes back into animal feed). Direct food uses comprise the other 11%. In 2007, when ethanol production consumed about a quarter of the US corn crop, the US had its largest corn export volumes ever, and we finished the year with unused corn stocks. The market has not been short on corn

Rising global food prices are a real and important concern, particularly in the developing world where food can be a very significant portion of a family budget. Agriculture and associated advances in processing technologies can supply the solutions, providing adequate resources for food, feed, materials, and fuel far into a sustainable future – just as they have done for as far back as we care to look. Slowing US biofuels production is not the solution. Let's not take actions that will do nothing to solve food price inflation but will certainly harm the advancement of next generation biofuels. Let us instead continue to expand agricultural production from the world's existing farm acreage, a goal we are already addressing and which we can readily accomplish.

Thank you for the opportunity to speak with you today on this important topic, and I look forward to your questions.

In: Renewable Fuel Standard Issues
Editor: Daniel T. Crowe

ISBN 978-1-60692-289-7

Chapter 8

TESTIMONY OF SCOTT FABER (FEDERAL AFFAIRS ON BEHALF OF THE GROCERY MANUFACTURERS ASSOCIATION) BEFORE THE SENATE COMMITTEE ON ENVIRONMENT AND PUBLIC WORKS ON "ENVIRONMENTAL PROTECTION AGENCY OVERSIGHT: IMPLEMENTING THE RENEWABLE FUEL STANDARD" JULY 10, 2008

Thank you for the opportunity to testify on the implementation of the Renewable Fuel Standard by the Environmental Protection Agency.

Food and beverage companies strongly support the development of sustainable bio-fuels that contribute to America's energy security and that do not pit the nation's energy needs against the needs of the hungry or the environment. We are working with a broad coalition of industry, farm, labor, anti-hunger, consumer, minority, and environmental organizations to urge Congress to revisit our food-to-fuel policies in light of runaway food inflation and new questions and concerns regarding the environmental costs of food-to-fuel production.

While there are many factors contributing to the sharp increase in US and global food prices – including poor weather, export restrictions, rising demand for food globally, the weak dollar, higher energy prices, and changes in commodities markets -- the most significant new factor and the only factor affecting food and feed prices that is under the control of the Congress, is the sudden and significant increase in food-to-fuel production.

I have attached analyses by the World Bank, [1] IMF, [2] UN FAO, [3] CRS, [4] USDA-ERS, [5] IFPRI, [6] Oxfam [7] and by former USDA Chief Economist Keith Collins [8] which document the combination of factors contributing to US and global food prices and the significant role of food-to-fuel production. In general, the rapid expansion of corn ethanol and biodiesel production has increased demand for corn and vegetable oil, increased the price of products which use corn and vegetable oil as ingredients, and increased the price of other crops that compete with corn and soybeans for land.

Commodity prices are rising at dramatic rates. Since the 2005 crop year, farm- level corn prices have increased more than 200 percent, and farm-level soybean prices have increased more than 135 percent. [9] Although there are many factors contributing to increases in commodity prices, the recent surge in ethanol production is one of the most significant factors. Between 2006 and 2008, US corn ethanol production accounted for 75 percent of the growth in global demand for coarse grain and 50 percent of the growth in demand for all grains. [10] Collins estimates that corn ethanol could be responsible for as much as 60 percent of the expected increase in corn prices between the 2006 and 2008 crop years. [11] As Collins notes, "the increase in corn demand due to ethanol is rising faster than growth in corn yields per acre. So long as that situation continues, corn will have to attract acreage from other crops to meet its expanding demand. This shift will mean higher prices for all crops that compete, directly or indirectly, for acreage with corn." [12]

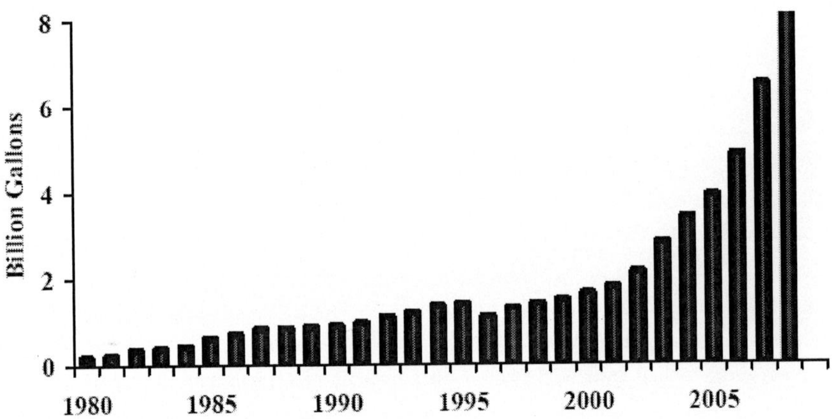

U.S. Ethanol Production 1980-2008.

Food prices are now rising at twice the overall rate of inflation. Because the price of basic commodities has dramatically increased, domestic food prices rose by 4.9 percent during 2007 – twice as fast as inflation and the largest increase in 17 years. Food prices for the previous three months have increased at a seasonally adjusted annualized rate of 6.3 percent, and studies predict that annual food price inflation will average 9 percent between 2008 and 2012 as the impact of rising commodity prices are slowly reflected in retail prices. [13] Although there are many factors contributing to food price inflation, the rising cost of commodities – driven in large measure by growing food-to-fuel production – is expected to cause food prices to rise 23 to 35 percent faster than historical increases in food prices. [14] In particular, the price of animal products will continue to rise dramatically in response to higher feed prices. [15] Between May 2005 and May 2008, food-to-fuel production contributed to increases in the costs of basic staples like eggs (62.8 percent), whole milk (17.2 percent), and whole chicken (13.5 percent). [16].

Rising food prices fall most heavily on the poor. These price increases fall most heavily on the poorest 20 percent of Americans who spend roughly one-third of their after-tax income on food and on the global poor who spend as much as 70 percent of their income on food. Rising commodity prices have pushed global food prices up 83 percent over the last three years [17] - and by 57 percent in the last year – pushing 50 million people into poverty in 2007 alone, according to the UN FAO. In combination, rising prices and declining commodity stocks have forced global food aid programs to ration food, and have contributed to food riots and protests in more than 30 countries. Rising food inflation in the developing world is not merely a food security issue, but is a national security issue. The World Bank warns that 33 nations are at risk of social unrest because of the rising price of food. [18]

Rising food prices pose significant budgetary challenges. Although potential outlays are difficult to estimate, government spending will increase significantly as food prices rise. Many federal programs linked entirely or in part to the Consumer Price Index (CPI), including anti-hunger assistance programs and child nutrition programs. The CPI is often used to adjust federal payments, determine program eligibility, and to provide cost-of-living adjustments to millions of workers. Overall, the CPI affects the income of about 80 million people, including 51.6 million Social Security beneficiaries, 21.3 million food stamp recipients, about 4.6 million military and civil service retirees or survivors, and more than 2 million workers impacted by collective bargaining agreements. In particular, changes in the CPI affect the cost of school lunches for 28.4 million children. [19] Rising food prices will impact federal outlays in three ways: by automatically increasing federal expenditures on programs linked all or in part to the CPI; by reducing the number of households and students that can be served by programs, such as the national school lunch program; and by forcing appropriators to reduce discretionary spending for other programs to address shortfalls.

Rising feed prices are causing severe economic harm for livestock producers. Although some crop farmers have benefited from high commodity prices, many more livestock producers are facing unprecedented losses. The higher costs of corn and soybean meal [20] have translated directly into higher feed costs for all livestock producers. [21] Feed costs climbed by over $15 billion between 2005 and 2008 due to higher prices for corn and other grains. [22] Moreover, feed costs will continue to remain well above historic levels through 2017 as food-to-fuel mandates are fully implemented. [23] In 2008-09, for example, food-to-fuel mandates are estimated to increase the cost of livestock production by as much as $17.7 billion. [24] Ultimately, the increased cost of feed will be passed on to consumers in the form of higher food prices. As feed costs rise, meat and poultry production will decline and many livestock farmers will be forced out of business. Producers at greatest risk of failure are poultry, dairy, hog and beef producers who do not produce their own feed grains. [25]

Historically low commodity stocks pose severe economic consequences. Global stocks of several major commodities are at or near historic lows – particularly when measured as a share of total usage. [26] For example, global end-of-year stocks for coarse grains and wheat are projected to drop by mid-2008 to the lowest levels since 1977, while ending stocks of total grains will fall to the lowest levels since 1981. In particular, a rapid increase in the production of ethanol combined with a decline in corn plantings will likely result in the second lowest level of corn stocks relative to consumption in 49 years. Increasing the use of corn for ethanol by 33 percent in 2008 will contribute to a 40 percent reduction in the corn inventory. [27] For most commodities, annual prices tend to have a strong negative correlation with the ending stocks-to-use ratio. [28] Reduced yields in 2008 caused by a wet spring and flooding combined with surging ethanol production and low commodity stocks are already resulting in dramatic increases in crop prices.

Food prices will continue to rise as more and more corn and soybean oil are diverted to our fuel supplies. Unless the Congress and the Administration act this year to revise federal food-to-fuel mandates, commodity prices will continue to rise as more and more food is diverted to our fuel supplies. In 2008, roughly one-third of U.S. corn supplies will be diverted to produce fuel. In the coming years, 40 percent or more of the U.S. corn crop and as much as 30 percent of U.S. vegetable oils will be diverted from our food supplies to our fuel supplies.

Because commodity prices will remain high in response to these mandates, food prices are expected to increase by 9 percent annually between 2008 and 2012. [29]

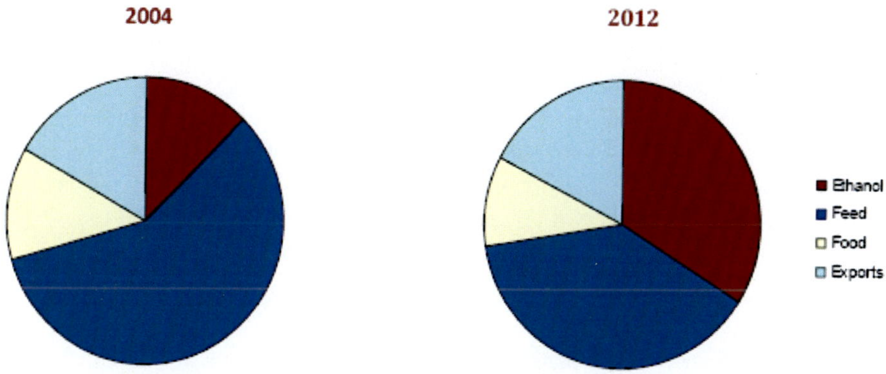

Because ethanol displaces a small fraction of the US gasoline supply and a tiny fraction of global crude supplies, food-to-fuel mandates currently have no impact on gasoline prices. Overall, ethanol production in 2007 displaced less than 4 percent of the nation's gasoline supplies in 2007, when relative energy values are considered. Consequently, freezing the mandate at the levels blended in 2007 – that is, reducing the mandate from 9 and 10.5 billion gallons to levels produced in 2007 – would not increase gasoline prices. In fact, failure to revise food-to-fuel mandates could marginally increase gasoline prices under some scenarios. Eventually, rising demand for corn to produce ethanol will increase the cost of producing ethanol and result in higher prices at the pump. [30]

Recommendations

We urge the Committee to revisit and restructure our food-to-fuel policies to accelerate the development of fuels that do not pit our energy needs against the needs of the hungry and the environment. In particular:

Congress should freeze food-to-fuel mandates. In light of crop reports and perilously low commodity stocks, Congress and the Administration should act now to reduce the federal food-to-fuel mandate for 2008 and 2009 to production levels for 2007 and should revisit and revise food-to-fuel mandates, subsidies and tariffs to reflect changing economic conditions and new questions regarding the economic and environmental costs of fuels made from food crops. Freezing the mandate would result in immediate reductions in the price of corn. A recent study by FAPRI estimated that implementation of the RFS increased corn prices by 19 percent. [31] A separate study by IFPRI concluded that a freeze of biofuel production at 2007 levels would reduce global corn prices. [32] What's more, these studies do not take into account significant declines in yields in 2008, which are amplifying the impact of increased ethanol production on corn prices.

Congress should carefully evaluate the environmental impacts of food-to-fuel policies. Diverting food crops to our fuel supplies has artificially increased the price of commodities, accelerating the conversion of pasture and forest lands to crop production at home and around the globe. Current and expected conversion of pasture and forest lands will release carbon into the atmosphere and reduce the availability of carbon "sinks" that help sequester carbon. In addition, increased production of row crops has increased water pollution, compounded water shortages, and contributed to the loss of habitat for wildlife. In particular, increased fertilizer use associated with expanded crop production has increased the amount of nitrogen and phosphorous being washed into rivers and bays, including the Chesapeake Bay and the Gulf of Mexico, and will increase ground-level ozone in some regions. Increasing the use of distiller's grain – a byproduct of ethanol production that is fed to animals but has less nutritional value than feed - increases the amount of phosphorous reaching waterways.

In particular, Congress should freeze food-to-fuel mandates to carefully assess the life-cycle emissions caused by bio-fuels. Reducing emissions from transportation fuels involves the consideration of numerous factors, including a common set of accounting principles and the ability to verify emissions reductions. Because the development of these accounting principles and verification methods are still underway, Congress should freeze our food-to-fuel mandates to ensure that EPA and other policymakers and experts have ample time and resources to adequately assess and verify potential emission reductions from bio-fuels. Significant questions regarding the life- cycle environmental effects of biofuels, including the significant effects of land cultivation, remain unresolved.

Congress should accelerate the development of advanced and cellulosic bio-fuels. Congress should revisit and reform food-to-fuel mandate schedules, subsidies and tariffs to gradually reduce our reliance on food as an energy feedstock. In particular, Congress should accelerate the development of cellulosic and advanced bio-fuels derived from fuel feedstocks that do not increase food or fiber prices and that improve the environment. Many of these fuels can be produced from feedstocks that do not compete with food crops, provide significant reductions in emissions when compared to gasoline, can be distributed through existing infrastructure, and could displace a significant share of our gasoline supplies. To accelerate the development of such fuels, Congress should eliminate the tariff on imported bio-fuels, should consider reforms to federal tax credits to reward the production of sustainable bio-fuels, and should adopt a technology neutral standard for life-cycle reductions in emissions that applies to all fuels, including all corn ethanol and bio-diesel produced regardless of the date of plant construction. Setting a lower bar for conventional bio-fuels – by setting lower emissions requirements or by exempting production from emissions requirements altogether – creates a competitive disadvantage for advanced and cellulosic biofuels that should be addressed.

Congress should accelerate global agricultural development. Congress should take steps to expand hunger assistance programs to help address the impact of food-tofuel policies on food inflation at home and abroad. What's more, Congress should also provide new funds to increase the productivity and sustainability of agricultural lands in the developing world. Between 2003 and 2007, global usage of coarse grains like corn grew by 3.4 percent. At the same time, annual growth in agricultural productivity is slowing. Between 1970 and 1990, production rose by an average of 2.2 percent per year. Since 1990, the growth rate has declined to about 1.3 percent. Projections for US and world agriculture see the rate declining to 1.2 percent per year between 2009 and 2017. [33]

In conclusion, we urge Congress to revisit the food-to-fuel policies in light of dramatic increases in food prices and new questions about the environmental costs of fuels derived from food crops. Although there are many factors contributing to record food inflation – including increasing global demand, export restrictions, poor weather, commodity speculation, and higher energy prices – a significant new factor and the only factor affecting food and feed prices that is under the control of Congress is food-to-fuel mandates and subsidies diverting food into our fuel supplies.

Reference

[1] "Rising Food Prices: Policy Options and World Bank Response," World Bank, April 2008.

[2] International Monetary Fund, "World Economic Outlook, Globalization and Inequality," October 2007.

[3] UN FAO, Soaring Food Prices: Facts, Perspectives, Impacts and Required Actions, June 2008.

[4] CRS, High Agricultural Prices: What are the Issues?, May 6, 2008.

[5] USDA, ERS, Global Agricultural Supply and Demand: Factors Contributing to The Recent Increase in Food Commodity Prices, May 2008.

[6] Von Braun, Joachim. "Biofuels, International Food Prices, and the Poor." International Food Policy Research Institute (IFPRI). June 12, 2008.

[7] "Another Inconvenient Truth: How biofuel policies are deepening poverty and accelerating climate change," Oxfam International, June 26, 2008.

[8] Collins, Keith. "The Role of Biofuels and Other Factors in Increasing Farm and Food Prices," prepared for Kraft Global Food, June 2008. (Hereinafter "Collins")

[9] USDA, World Agricultural Supply and Demand Estimates, June 2008.

[10] Id.

[11] 11 Collins. Collins finds that ethanol could account for 60 percent of the expected increase in corn prices between 2006/07 and 2008/09 when market demand and supply are inelastic with respect to price – that is, a period when stocks are very low, feed use is slow to respond, export demand is strong due to foreign agricultural policies, and acreage is very constrained.

[12] Id.

[13] Advanced Economic Solutions, Rising Commodity Prices and their Impact on US Food Inflation, June 2008.

[14] Collins. If food-to-fuel production accounts for 60 percent of the expected increased in feed grain and oilseed product costs between 2006/07 and 2008/09, and these increases are passed on to retail consumers, these increases would increase baseline US expenditures on food by 1.8 percent over a 2-3 year period. This increase is significant in light of the fact that long-term annual average food inflation is about 2.5 percent. Thus, the increase in retail food prices due to biofuels is estimated to be 23-35 percent above the normal increase in food prices that would occur over 2-3 years.

[15] Elam, Thomas, "Biofuels Support Policy Costs to the U.S. Economy," FarmEcon LLC, March 24, 2008 (Hereinafter "Elam")

[16] Consumer Price Index – Average Price Data (retrieved for most requested statistics), Bureau of Labor Statistics, available at http://www.bls.gov/data/home.htm.

[17] Bob Davis and Douglas Belkin, Food Inflation, Riots Spark Worries for World Leaders," *Wall Street Journal*, April 14, 2008. A1.

[18] "The World Food Crisis," New York Times, Editorial, April 10, 2008.

[19] BLS, at http://www.bls.gov/dolfaq/bls_ques1.htm

[20] Elevated corn prices in response to increasing biofuels demand also contribute to the historically high prices of soybeans because soybean cropland must compete with corn for cropland. Indeed, the price of

soybeans has risen even though stockpiles of soybean oil are also at near record levels. USDA reported that the price of soybeans per bushel was $6.43 in 2006/2007, but shot up to $9.00 in 2007/2008. USDA Projections, at 35 (Table 7). Prices of soybeans and soybean-derived products (e.g., soybean meal and soybean oil) are projected to increase and remain well over 2006/2007 levels over the long-term due to continued increased demand for biofuels, with even greater price increases likely as a result of the strengthened biofuels mandates enacted by Congress in 2007. USDA Projection, at 23 and 41 (Table 13).

[21] Food and Agricultural Policy Research Institute, 2008 U.S. Baseline Briefing Book (March 2008) (hereinafter "FAPRI Report"), at 3, available at http://www.fapri.missouri.edu/outreach/publications/2008/FAPRI_MU_Report_03_08.pdf.

[22] FAPRI Report, at 60.

[23] FAPRI Report, at 60. See also USDA Projections, at 4, 49 and 60. In fact, USDA acknowledges that its own projections likely underestimate the anticipated increases in costs of animal feed because, although they account for the biofuels mandates created by the Energy Policy Act of 2005, they do not account for the strengthening of those mandates by Congress in December 2007, which has added to the unprecedented demand for corn. USDA Projections, at 23.

[24] Elam at 28,

[25] Elam at 28. See also FAPRI Report, at 42 (suggesting that beef producers will experience financially difficult times in the next few years as they face high and rising input costs); see also FAPRI Report, at 50 (stating that higher feed costs have contributed to increases in the costs of producing milk).

[26] CRS, High Agricultural Prices: What are the Issues?, May 6, 2008.

[27] USDA-ERS, Feed Outlook, May 13, 2008.

[28] CRS, High Agricultural Prices: What are the Issues?, May 6, 2008.

[29] 29 Advanced Economic Solutions, The Impact of Rising Commodity Prices on Food Inflation, June 2008

[30] Elam, Thomas, "Biofuels Support Policy Costs to the U.S. Economy," FarmEcon LLC, March 24, 2008.

[31] FAPRI-MU, The Energy Independence and Security Act of 2007: Preliminary Evaluation of Selected Provisions, January 2008.

[32] Rosegrant, Biofuels and Grain Prices: Impacts and Policy Responses, May 5, 2008.

[33] USDA, ERS, Global Agricultural Supply and Demand: Factors Contributing to The Recent Increase in Food Commodity Prices, May 2008.

INDEX

A

absorption, 127
academic, 38
academics, 129
accountability, 91
accounting, 4, 8, 20, 114, 152
achievement, 48
acid, 32, 49
acute, 62, 138
adaptability, 115
addiction, 4, 23
additives, 76
administrative, 83
Administrative Procedure Act, 57
Advanced Energy Initiative, 29, 30, 35, 47, 65
Advanced Energy Initiative (AEI), 29, 35, 47
AEI, 29, 47, 119, 123
Africa, 50, 140
Ag, 119, 121, 122, 123
agents, 19
agricultural, vii, 8, 9, 17, 18, 46, 89, 98, 99, 100, 101, 102, 104, 106, 108, 110, 111, 115, 118, 120, 126, 129, 132, 134, 138, 139, 140, 141, 143, 144, 145, 153, 154
agricultural commodities, 138, 143
agricultural crop, 110
agricultural market, 99, 111
agricultural residue, 17, 18, 100, 120, 134
agricultural sector, 115, 129
agriculture, 5, 7, 9, 10, 19, 37, 111, 139, 140, 141, 144, 153
air, vii, 2, 27, 29, 32, 38, 40, 46, 47, 56, 62, 63, 77, 92, 117, 129, 134, 138
Air Force, 41, 53
air pollutant, 38, 46, 47, 62
air pollutants, 38, 46, 47
air quality, 2, 56, 92, 117, 129, 134, 138
air toxics, 47
air traffic, 40
Alaska, 88
alcohol, 34, 43, 53, 138
alcohols, 101
algae, 16, 75, 109, 134
alternative, vii, viii, 15, 23, 30, 39, 43, 47, 53, 55, 56, 74, 97, 98, 99, 102, 105, 115, 118, 131, 138, 141
alternative energy, 56, 102, 118
alternatives, 39
Amazon, 24
Amazonian, 8
ambient air, 62
amendments, 50, 72, 73, 95
American Competitiveness Initiative, 35, 66
American Competitiveness Initiative (ACI), 35
American Jobs Creation Act, 48
AMS, 19, 112
analysts, 70
analytical tools, 133
animals, 113, 152

Animals, 11
annual rate, 111
antenna, 39
AOC, 81
APP, 36
application, 38, 84, 85, 86
appropriations, 34, 35, 37, 42, 48, 53, 60, 86, 134, 135
appropriations bills, 37, 48
Appropriations Committee, 29, 30, 36, 49
Arctic, 66, 67
Arctic National Wildlife Refuge, 66, 67
argument, 13, 107
Arizona, 25
Army, 38, 55, 56
Army Corps of Engineers, 38, 55, 56
Asia, 36, 37, 50, 140, 143
assessment, 17, 25, 37, 39, 82, 87, 88, 89, 141
assessment tools, 141
assumptions, 13, 15, 129
Atlantic, 38
atmosphere, viii, 32, 125, 152
Atomic Energy Act, 91
attribution, 144
auditing, 19, 21
Australia, 135
authority, 20, 21, 39, 55, 56, 75, 88, 132
automakers, 117
availability, vii, 31, 41, 69, 70, 89, 115, 117, 119, 130, 133, 152
awareness, 77

B

background information, 58
bacterial, 16
barges, 116
barrier, 21, 45, 94, 100, 105, 118
barriers, 93, 94, 98
batteries, 63
battery, 71, 74, 77
beef, 8, 113, 130, 150, 155
behavior, 25
Beijing, 3
benchmarks, 134
benefits, 2, 3, 5, 8, 9, 22, 23, 38, 44, 56, 80, 102, 107, 108, 110, 115, 118, 120, 142
biodegradable, 47, 102
biodiesel, 1, 4, 7, 23, 42, 43, 47, 48, 52, 53, 62, 63, 75, 98, 99, 101, 106, 108, 109, 118, 135, 148
biodiversity, 11, 12, 13
bioethanol, 46
biofuel, vii, viii, 4, 6, 7, 21, 43, 75, 97, 98, 99, 100, 101, 102, 103, 105, 106, 107, 109, 110, 115, 116, 117, 118, 120, 128, 135, 139, 152, 154
biofuels, vii, viii, 1, 2, 3, 4, 5, 6, 7, 8, 9, 10, 12, 13, 14, 15, 16, 17, 18, 20, 21, 22, 23, 24, 27, 31, 47, 49, 52, 61, 71, 75, 76, 91, 97, 98, 99, 100, 101, 102, 105, 106, 107, 108, 109, 110, 111, 115, 116, 118, 119, 127, 131, 132, 133, 134, 135, 136, 137, 138, 139, 141, 142, 143, 144, 152, 154, 155
biogas, 75
biomass, 2, 3, 4, 5, 6, 7, 9, 10, 11, 12, 13, 14, 16, 17, 18, 19, 21, 25, 29, 31, 34, 36, 41, 42, 43, 45, 46, 47, 49, 55, 60, 61, 63, 75, 87, 98, 100, 101, 103, 106, 107, 110, 116, 120, 127, 128, 132, 134, 135, 136, 138
biomaterials, 137, 138
biorefinery, 47, 138, 139, 142
biotechnology, 30, 47
bipartisan, 73
birds, 12, 39
blends, 60, 76, 117, 118, 127, 138, 139
boats, 38
bogs, 143
boilers, 71, 77, 81
bonds, 52
Botanical Garden, 24
bounds, 140
Brazil, 8, 60, 100, 101, 109, 117, 119, 140
Brazilian, 8, 99, 100, 106, 118
breeding, 47
broilers, 112
buildings, 24, 37, 77, 78, 80, 82, 91, 93
bulbs, 71, 77, 82
Bureau of Land Management, 66
burn, 26

burning, 7, 8, 14, 139
burns, 14
buses, 53
Bush Administration, 29, 37
bust, 14

C

Canada, 47, 92, 100, 105, 135
cancer, 62
cane sugar, 60
capacity, 16, 17, 35, 38, 45, 46, 59, 62, 87, 88, 89, 104, 105, 110, 114, 115, 117, 119, 132
carbon, vii, 1, 3, 7, 9, 11, 13, 14, 15, 17, 18, 22, 23, 46, 47, 48, 49, 62, 81, 87, 88, 102, 110, 133, 138, 139, 141, 142, 143, 152
Carbon, 7, 22, 23, 24, 87
carbon credits, 102
carbon dioxide, vii, 1, 3, 46, 48, 49, 81, 87, 88
carbon emissions, 81, 142
carbon monoxide, 47, 62
Caribbean, 100, 119
Caribbean Basin Initiative, 100, 119
Caribbean Basin Initiative (CBI), 100
Caribbean countries, 100
cartel, 33
cattle, 8, 113, 130
cell, 46, 87
cellulose, 16, 17, 46, 47, 98, 101, 106, 138
Cellulose, 51, 100
cellulosic, 2, 4, 6, 15, 16, 17, 30, 36, 42, 46, 47, 52, 58, 61, 71, 75, 76, 98, 99, 101, 103, 105, 106, 107, 108, 109, 110, 119, 127, 128, 132, 134, 135, 138, 139, 141, 142, 152
cellulosic ethanol, 4, 30, 36, 46, 47, 61, 71, 75, 76, 105, 108, 119, 127, 132, 134, 135, 138, 141
cement, 144
certification, 13, 19, 21
Chad, 121
channels, 111
chemicals, 106, 140
chicken, 149
child nutrition programs, 149
childhood, 94
children, 149
China, 111, 114, 135
civil service, 149
civilian, 39
Clean Air Act, 24, 66, 76, 132
clean energy, 50, 79, 90
Clean Renewable Energy Bond, 43
Clean Renewable Energy Bonds, 43
clean-energy, vii, 29
climate change, 32, 81, 91, 154
Clinton Administration, 33
closed-loop, 34, 41, 55
cloture, 73
clouds, 142
CO2, 8, 9, 20, 48, 49, 63, 143
coal, 23, 81, 86, 87, 99, 110
Coast Guard, 30, 39, 57, 82
coastal areas, 38
collaboration, 126
collective bargaining, 149
Colorado, 26
Columbia, 41, 82
combustion, 2, 5, 31, 34, 49, 79, 81, 107
commerce, 64, 76
commercialization, 16, 33
Committee on Appropriations, 30, 49, 50
Committee on Environment and Public Works, 1, 125, 131, 137, 147
commodity, 13, 17, 102, 112, 114, 118, 144, 148, 149, 150, 151, 153
Commodity Credit Corporation, 53
communities, 11
community, 14, 126, 129
compaction, 13
compatibility, 141
compensation, 90
competition, 2, 8, 10, 59, 113, 139
competitiveness, vii, 29
complexity, 22, 126, 129
compliance, 19, 21, 74, 83, 127, 128, 130
components, 84, 103, 116
concentrates, 113
concentration, 113
conditioning, 77, 139
Conference Committee, 53

confidence, 22
Congress, viii, 1, 2, 3, 5, 7, 18, 20, 22, 23, 30, 33, 34, 35, 39, 44, 49, 53, 55, 59, 60, 63, 64, 66, 67, 70, 72, 75, 82, 85, 86, 88, 89, 90, 91, 92, 93, 95, 96, 118, 128, 130, 131, 132, 133, 141, 142, 143, 147, 148, 150, 151, 152, 153, 155
conifer, 15
conservation, 5, 13, 17, 21, 83, 118, 133
constraints, 89, 101, 109
construction, 18, 38, 40, 79, 81, 91, 92, 94, 104, 110, 115, 128, 153
construction and demolition, 18
consulting, 130
Consumer Price Index, 111, 149, 154
consumers, 44, 88, 114, 138, 150, 154
consumption, 33, 61, 74, 98, 99, 101, 107, 108, 113, 114, 117, 118, 120, 129, 143, 150
contracts, 104, 127
control, 26, 40, 48, 79, 139, 148, 153
conversion, 2, 6, 8, 9, 12, 13, 16, 18, 19, 31, 47, 75, 105, 107, 120, 135, 152
conversion rate, 18, 120
cooking, 11, 62, 101
cooling, 78
COP, 67
Coping, 25
copper, 144
corn, viii, 4, 6, 8, 9, 15, 16, 30, 45, 46, 47, 75, 97, 98, 99, 100, 101, 102, 103, 104, 105, 106, 107, 108, 109, 110, 111, 112, 113, 114, 115, 116, 118, 119, 121, 123, 130, 132, 134, 136, 137, 139, 140, 144, 148, 150, 151, 153, 154, 155
corn-starch ethanol, 105
corporate average fuel economy, 72
Corporate Average Fuel Economy, 64, 69, 70, 73
correlation, 150
corrosion, 116
corrosive, 116
corruption, 91
cost saving, 81, 86, 106
cost-effective, 74, 79, 83, 135
cost-of-living adjustments, 149
costs, viii, 8, 14, 25, 30, 32, 35, 47, 73, 80, 81, 82, 83, 92, 95, 99, 102, 105, 106, 109, 111, 113, 114, 116, 118, 120, 123, 131, 138, 144, 147, 149, 150, 151, 153, 154, 155
costs of production, 106
covering, 52, 56
CPI, 111, 121, 149
credit, 1, 4, 18, 23, 34, 35, 41, 48, 52, 55, 62, 128
criticism, 102
crop production, 109, 110, 152
crop residues, 10, 120, 129
crops, 2, 7, 8, 9, 10, 17, 45, 46, 49, 55, 98, 99, 100, 106, 107, 110, 111, 120, 129, 133, 134, 144, 148, 151, 152, 153
CRS, 29, 31, 40, 49, 55, 59, 60, 61, 63, 64, 66, 67, 69, 70, 71, 72, 74, 75, 94, 96, 97, 100, 112, 119, 120, 121, 148, 153, 155
crude oil, 89, 121
cultivation, 2, 5, 19, 110, 152
customers, 44, 59, 96
cycling, 133
cyclists, 92

D

dairy, 113, 150
data collection, 89
database, 122
debentures, 93
Defense Authorization Act, 39, 40
Deficit Reduction Act, 53
definition, 6, 7, 9, 10, 11, 12, 19, 20, 21, 24, 55, 81, 92, 96, 109, 118, 120, 128, 129
deforestation, 12, 49
degradation, 2, 4, 6, 118
delivery, 7, 76, 83, 107, 144
Department of Agriculture, 37, 132
Department of Commerce, 90
Department of Defense, 30, 36, 39, 50, 53, 82
Department of Defense (DOD), 30, 39, 82
Department of Education, 79
Department of Energy, 29, 30, 65, 68, 119, 120, 131, 136
Department of Energy (DOE), 30, 120

Department of Homeland Security, 40
Department of Housing and Urban Development, 80
Department of Housing and Urban Development (HUD), 80
Department of Interior, 30, 34, 38, 55
Department of Interior (DOI), 30
Department of State, 32, 37, 50, 53, 54, 63, 90
Department of the Interior, 66, 87, 89
Department of Transportation (DOT), 91, 92
deposits, 96
developing countries, 37, 50, 53, 90, 114, 135
Development Assistance, 50
diesel, 4, 6, 7, 9, 15, 16, 47, 62, 75, 99, 101, 109, 116, 117, 127, 128, 132, 135, 153
diesel engines, 47
diesel fuel, 16, 47, 48, 62, 101, 127, 128
differentiation, 142
directives, 35, 36
disaster, 130
discounting, 143
discretionary, 150
discretionary spending, 150
displacement, 49, 134
distortions, 113
distributed generation, 44
distribution, 7, 37, 44, 47, 85, 86, 93, 116, 127, 128, 130
District of Columbia, 41, 82
diversity, 10, 12, 13, 26
draft, 20, 38, 56, 129
drought, 31, 108, 140
droughts, 140
drowning, 94
DuPont, viii, 137, 139
DuPont de Nemours, 137
duration, 81
duty free, 100

E

East Asia, 50, 140
ecological, 11, 12, 13, 25, 26
ecological restoration, 25, 26
economic activity, 115
economic growth, 111
economic incentives, 109
Economic Research Service, 122
economic security, 131
economics, 9, 109, 111, 141
economies of scale, 113
ecosystem, 12, 13, 39
ecosystems, 2, 5, 7, 11, 17, 87, 143
Education, 79
egg, 114
electric energy, 59
electric power, 33, 43, 91
electric power production, 43
electric utilities, 35, 59, 71
electricity, vii, 14, 29, 31, 32, 34, 37, 41, 44, 45, 48, 59, 71, 72, 73, 82, 83, 85, 94, 106
eligible countries, 91
embargo, 33, 107
emission, 3, 8, 49, 63, 136, 152
employers, 95
employment, 111, 115
encouragement, 142
energy, vii, viii, 2, 3, 14, 15, 22, 23, 25, 27, 29, 30, 31, 32, 33, 34, 35, 36, 37, 38, 39, 41, 42, 43, 44, 45, 46, 47, 48, 49, 50, 52, 53, 54, 55, 56, 57, 59, 60, 61, 62, 63, 64, 66, 67, 68, 69, 70, 71, 72, 73, 77, 78, 79, 80, 81, 82, 83, 84, 85, 86, 87, 88, 89, 90, 91, 93, 98, 99, 101, 102, 104, 106, 107, 108, 109, 111, 116, 118, 120, 123, 125, 129, 131, 133, 134, 135, 136, 138, 139, 141, 144, 147, 151, 152, 153
Energy and Water Development, 34, 36, 37, 50
energy audit, 81, 83
energy consumption, 33, 107, 108
energy density, 138, 139, 141
energy efficiency, vii, 32, 33, 35, 37, 48, 49, 50, 53, 69, 70, 71, 72, 73, 77, 78, 79, 80, 81, 82, 83, 86, 90, 93
Energy Efficiency and Renewable Energy, 30, 31, 35, 44, 49, 55, 59, 60, 63, 64, 65, 67, 68, 96, 136
Energy Efficiency and Renewable Energy (EERE), 30, 35

Energy Independence and Security Act, vii, 2, 10, 24, 69, 70, 96, 125, 131, 155
Energy Information Administration, 31, 54, 65, 88, 89, 120
Energy Information Administration (EIA), 88, 89, 120
energy markets, 90
Energy Policy Act, 30, 34, 38, 41, 42, 52, 76, 78, 101, 104, 125, 155
Energy Policy Act of 2005, 30, 34, 41, 42, 52, 76, 78, 101, 104, 125, 155
energy recovery, 79
energy supply, vii, 33, 98, 99, 101
engagement, 129
engines, 47
enterprise, 138
entrapment, 94
environment, viii, 3, 5, 38, 58, 87, 99, 132, 147, 151, 152
environmental degradation, 4, 118
environmental effects, 98, 106, 152
environmental impact, 2, 4, 5, 30, 38, 39, 44, 105, 133, 152
Environmental Protection Agency, 48, 62, 66, 68, 76, 125, 132, 147
environmental standards, 56
enzymatic, 105
enzymes, 16, 47
EPA, 2, 3, 16, 17, 18, 19, 20, 21, 22, 24, 48, 62, 75, 76, 78, 79, 80, 87, 92, 117, 121, 126, 128, 129, 130, 132, 133, 143, 152
epoxy, 116
erosion, 18, 139
ethanol, viii, 1, 4, 6, 8, 9, 15, 16, 18, 23, 30, 34, 36, 42, 45, 46, 47, 52, 58, 60, 61, 71, 75, 76, 81, 97, 98, 99, 100, 101, 102, 103, 104, 105, 106, 107, 108, 109, 110, 111, 112, 113, 114, 115, 116, 117, 118, 119, 120, 121, 122, 123, 127, 130, 132, 134, 135, 136, 137, 138, 139, 141, 144, 148, 150, 151, 152, 153, 154
Europe, vii, 29, 38
European Commission, 64
European Union, 32, 64, 135
excise tax, 1, 4, 5, 34, 48, 53
exclusion, 41
Executive Office of the President, 66
expansions, 110
expenditures, 114, 149, 154
expertise, 15
exploitation, 15
export promotion, 90
exporter, 114
exports, 114, 144
extraction, 7, 14, 128

F

FAA, 40
failure, 33, 150, 151
family, 144
family budget, 144
FAO, 148, 149, 153
Farm Bill, 21, 142
farmers, viii, 17, 112, 115, 130, 137, 150
farming, 8, 140, 143
farms, 8, 12, 38, 39, 40, 41
fats, 47
February, 24, 35, 52, 53, 54, 58, 60, 66, 112, 117, 119, 121, 122, 123, 126, 135
Federal Aviation Administration, 40
Federal Energy Regulatory Commission, 35, 65, 82, 89
federal government, 33, 48, 80
Federal Highway Administration, 123
federal law, 6
Federal Register, 56, 57, 126
Federal Trade Commission, 89
feeding, 44
feedstock, 7, 14, 16, 17, 19, 75, 98, 106, 108, 109, 111, 116, 117, 128, 130, 138, 141, 142, 152
feet, 26, 38, 45, 79, 108
fencing, 94
FERC, 35, 82
fermentation, 139
fertility, 107
fertilizer, 18, 45, 46, 108, 111, 132, 140, 144, 152
fertilizers, 106, 107

fiber, 7, 17, 46, 141, 152
field trials, 133
field-tested, 41
filtration, 12
finance, 80
financing, 81, 142
fire, 13, 14, 15, 25, 26
fires, 15, 26
firms, 19
first generation, 21
fishing, 38, 39
fission, 33
flexibility, 13, 81, 94, 96, 114
flight, 41
flooding, 132, 150
flow, 16, 93
fluorescent lamps, 71, 77
food, vii, viii, 1, 2, 4, 5, 7, 8, 9, 11, 17, 98, 99, 100, 102, 104, 106, 111, 113, 114, 130, 133, 134, 135, 138, 139, 141, 144, 147, 148, 149, 150, 151, 152, 153, 154
food aid, 149
food production, 10
food products, 111, 113
food stamp, 114, 149
forest management, 11
forest resources, 13, 134
Forest Service, 25, 26, 27
forestry, 13, 46, 89, 132
forests, 1, 2, 3, 6, 8, 10, 11, 12, 13, 14, 15, 17, 25, 26, 135, 143
fossil, vii, 19, 29, 32, 33, 44, 48, 49, 63, 75, 78, 102, 106, 108, 118, 134, 139, 142, 144
fossil fuel, vii, 19, 29, 32, 44, 48, 49, 63, 102, 106, 108, 139, 144
fossil fuels, 19, 48, 102, 140
fractures, 84
freezing, 151
fuel, vii, viii, 3, 5, 6, 7, 8, 9, 10, 14, 15, 19, 20, 22, 23, 25, 26, 29, 32, 37, 42, 43, 44, 45, 46, 47, 48, 49, 53, 55, 63, 70, 71, 72, 73, 74, 75, 76, 78, 81, 86, 87, 92, 97, 98, 99, 100, 101, 102, 103, 104, 105, 106, 107, 108, 109, 110, 116, 117, 118, 121, 123, 125, 126, 127, 128, 129, 130, 131, 132, 133, 134, 135, 138, 139, 141, 142, 143, 144, 147, 148, 149, 150, 151, 152, 153, 154
fuel cell, 43, 44, 53, 63, 87
fuel cycle, 107
fuel efficiency, 74, 86
fuel type, 141
funding, 29, 30, 33, 35, 36, 37, 47, 48, 49, 50, 52, 53, 76, 77, 80, 81, 82, 83, 86, 88, 91, 93, 94, 119, 133
funds, 35, 36, 48, 81, 83, 86, 88, 90, 94, 95, 115, 153
fusion, 33
futures, 104

G

gas, vii, 6, 7, 9, 22, 23, 29, 33, 42, 44, 45, 49, 55, 58, 70, 71, 72, 73, 74, 82, 84, 91, 96, 99, 102, 108, 109, 110, 128
gases, viii, 125
gasification, 135
gasoline, vii, viii, 2, 4, 5, 6, 9, 16, 20, 29, 30, 34, 42, 45, 46, 60, 61, 62, 75, 97, 98, 101, 103, 104, 105, 106, 108, 109, 110, 116, 117, 118, 120, 123, 126, 127, 134, 135, 138, 139, 141, 142, 151, 152
GDP, 115
gene, 62
general fund, 74
generation, 7, 16, 21, 36, 44, 45, 47, 59, 84, 88, 128, 135
generators, 44
genetic diversity, 13
Geneva, 119, 123
geophysical, 95
Georgia, 61, 135
geothermal, 31, 34, 41, 42, 43, 84, 85, 87, 89, 96
Germany, 38
GHG, 2, 4, 6, 7, 9, 10, 14, 15, 17, 18, 19, 20, 21, 22, 23, 82, 90, 102, 110, 128, 129, 134
global climate change, vii, viii, 29, 48, 97, 98, 125
global demand, 143, 148, 153

global warming, 2, 3, 6, 7, 8, 9, 11, 12, 13, 15, 23, 56
Globalization, 122, 153
goals, 3, 48, 82, 83, 87, 97, 98, 118, 125, 142
government, 6, 9, 16, 33, 35, 42, 48, 79, 80, 88, 92, 104, 105, 114, 115, 149
Government Accountability Office, 66, 78
Government Accountability Office (GAO), 66, 78
grain, 98, 99, 101, 111, 114, 115, 130, 139, 141, 148, 152, 154
grains, 99, 113, 115, 133, 134, 144, 148, 150, 153
grants, 37, 53, 55, 74, 76, 79, 80, 83, 84, 85, 90, 92, 93
grass, 46
grasses, 46, 99, 106
grasslands, 2, 5, 10, 11
grazing, 26
Great Lakes, 92
green buildings, 78, 80
greenhouse, viii, 2, 3, 5, 6, 7, 9, 17, 20, 21, 22, 23, 37, 38, 48, 49, 53, 60, 74, 75, 102, 103, 106, 107, 109, 110, 125, 128, 129, 131, 132, 133, 134, 136
Greenhouse, 7, 24, 110, 121
greenhouse gas, viii, 2, 3, 5, 6, 7, 9, 17, 20, 21, 22, 23, 37, 38, 48, 49, 53, 60, 74, 75, 102, 103, 106, 107, 109, 110, 125, 128, 129, 131, 132, 133, 134, 136
greenhouse gas (GHG), 75, 110, 128
greenhouse gases, viii, 7, 38, 125, 129
groups, 38, 39, 49, 114, 119, 126, 129
growth, 2, 5, 9, 10, 11, 13, 17, 43, 49, 91, 99, 104, 108, 111, 115, 132, 143, 144, 148, 153
growth rate, 153
GSA, 77, 78, 80, 82
guidance, 20
guidelines, 1, 3, 79, 80, 86, 94
Gulf of Mexico, 152

H

habitat, 11, 13, 18, 39, 143, 152
halogen, 86
harm, 132, 144, 150
harmful effects, 111
Harvard, 45, 60
harvest, 13, 14, 17, 104, 106, 133, 139, 142
harvesting, 11, 17, 19, 106, 107, 142
hazards, 39
health, 13, 62, 80
health effects, 62
hearing, 41, 127, 135
Heart, 25
heat, vii, 1, 3, 14, 31, 44, 46, 47, 61, 62, 79, 81, 94
heat pumps, 79
heating, 31, 77, 78, 102, 103, 135
heating oil, 102
higher education, 84, 85
highways, 8
hiring, 94
hog, 113, 150
hogs, 113
home heating oil, 103, 135
Homeland Security, 40, 44, 59, 66
horizon, 82, 142
horses, 139
hot water, 82, 96
House, 29, 30, 35, 36, 37, 41, 42, 43, 49, 50, 51, 71, 72, 73, 81, 96, 121
House Appropriations Committee, 35
household, 114
households, 114, 150
housing, 12, 78, 80
hub, 38
human, 13, 31, 48, 143
humanitarian, 114
humans, 62
hybrid, 44, 74, 92, 134
hybrids, 37
hydro, 34, 47
hydrocarbon, 142
hydrocarbon fuels, 142
hydrocarbons, 47
hydrogen, 36, 42, 44, 52, 86
hydrologic, 13
hydropower, 31, 41, 43, 50, 52

I

Idaho, 57
identification, 88
IEA, 68
Illinois, 40, 44
images, 25
imbalances, 138
IMF, 148
impact analysis, 129
implementation, viii, 3, 22, 32, 35, 55, 78, 90, 93, 95, 102, 125, 126, 128, 130, 131, 132, 142, 147, 152
importer, 108
imports, 36, 100, 102, 106, 107, 108, 109, 114, 117, 119, 121, 132
incandescent, 71, 77, 82, 86
incandescent light, 82, 86
incentive, 4, 6, 8, 10, 17, 23, 36, 42, 43, 52, 58, 59, 60, 81, 109, 111
incentives, 1, 4, 5, 17, 22, 23, 46, 55, 70, 72, 73, 74, 83, 104, 105, 119, 141
incidence, 25
income, 34, 39, 76, 114, 115, 144, 149
income tax, 34
incomes, 111, 115
independence, 120
India, 63, 111, 135
Indian, 43, 88
indication, 20
indicators, 134
indirect effect, 143
Indonesia, 7
industrial, 15, 46, 79, 139
industry, vii, viii, 4, 6, 11, 12, 13, 14, 16, 22, 23, 35, 38, 40, 43, 58, 60, 80, 82, 84, 94, 97, 98, 104, 115, 116, 118, 119, 126, 132, 138, 142, 147
inelastic, 154
infinite, 20
inflation, viii, 34, 35, 111, 144, 147, 149, 153, 154
inflationary pressures, 138
infrastructure, vii, 24, 42, 76, 86, 91, 93, 96, 98, 101, 102, 105, 116, 117, 127, 132, 138, 139, 141, 142, 152
injection, 87
innovation, 2, 17, 20, 139
Innovation, 86
insects, 140
Inspector General, 36
institutions, 80, 85, 141
institutions of higher education, 85
insulation, 86
integration, 83, 134
intensity, 22, 26, 49, 138
interaction, 26
interdisciplinary, 87
interference, 36
Intergovernmental Panel on Climate Change, 48
Internal Revenue Code, 95
International Energy Agency, 64, 68
international markets, 114
International Monetary Fund, 122, 153
invasive, 13, 107
invasive species, 13, 107
inventories, 114
investment, 16, 34, 35, 44, 52, 80, 93, 102, 116, 135, 141
investment capital, 102
investors, 16, 132
Ireland, 38
irrigation, 41
Israel, 90

J

Japan, vii, 29, 115
jet fuel, 127
job training, 91
jobs, 38, 115
Joint Committee on Taxation, 43
jurisdiction, 55
justification, 36

K

K-12, 80

L

labeling, 71, 77
labor, viii, 38, 147
Lafayette, 119
land, 2, 5, 7, 8, 9, 10, 17, 19, 20, 21, 25, 26, 38, 39, 88, 99, 106, 110, 128, 129, 140, 143, 148, 152
land use, 2, 5, 7, 8, 9, 10, 21, 110, 129, 140, 143
landfill, 4, 34, 41
landfill gas, 34, 41
landfills, 42
land-use, 7
language, 37, 49, 57
large-scale, 87, 111, 113, 133
law, vii, 2, 3, 6, 7, 24, 40, 41, 42, 49, 52, 53, 69, 70, 71, 73, 74, 77, 79, 85, 89, 94, 101, 126, 128, 129, 132
laws, 9, 53, 56, 63, 94
lawyers, 3
layoffs, 36
leadership, 135
learning, 54, 84
LED, 86
legislation, 3, 55, 72, 88, 96, 118, 126, 127, 128, 130
legislative proposals, 99
lenders, 104
life cycle, 75, 82, 133, 141, 142, 143
lifecycle, 2, 4, 5, 6, 7, 8, 9, 10, 17, 19, 20, 21, 22, 23, 78, 82, 109, 110, 110, 121, 128, 129, 131, 133, 134, 141, 142, 143, 152, 153
life-cycle cost, 78, 82
light trucks, 69, 70, 117
lignin, 47
likelihood, 40, 105
limitations, 74, 128, 140
liquor, 45
livestock, 8, 99, 113, 133, 150
Livestock, 122
loan guarantees, 52, 142
loans, 37, 80, 93
lobby, 8
local government, 35, 42, 79, 80, 88, 92
location, 14, 41, 55, 106
logging, 10, 12, 13, 14, 25, 26, 143
long distance, 113
Los Angeles, 3
losses, 150
lower-income, 114
low-income, 76, 114

M

machinery, 107
Madison, 135
magnet, 37
maintenance, 47, 79
Maintenance, 53
Malaysia, 7
management, 1, 5, 11, 19, 36, 87, 88
management practices, 87
mandates, 22, 150, 151, 152, 153, 155
manufactured housing, 78
manufacturing, 86, 138
maritime, 39
market, 2, 6, 8, 15, 16, 17, 19, 20, 21, 31, 33, 36, 48, 50, 59, 75, 76, 77, 88, 89, 98, 99, 100, 101, 102, 113, 114, 115, 117, 122, 127, 135, 139, 141, 142, 143, 144, 154
market economics, 8
market incentives, 33, 141
market penetration, 31, 48, 76
market prices, 114
market share, 31
market value, 17, 88, 141
marketing, 44, 104, 105, 111, 114
marketplace, 111, 113
markets, 21, 35, 59, 90, 102, 104, 106, 113, 114, 115, 127, 144, 147
Massachusetts, 3, 22, 30, 38, 44, 57
measurement, 93
measures, 44, 48, 71, 73, 77, 81, 90, 93
meat, 113, 150

mechanical energy, 31
media, 56, 58, 88, 121, 143
methane, 45, 87
methanol, 101
metric, viii, 125
Mexico, 25, 36
MGMT, 51
microbes, 46
microbial, 46, 105
microorganisms, 16
migration, 11
military, 38, 39, 40, 149
milk, 112, 114, 149, 155
million barrels per day, 46, 61, 107
mineral resources, 88
minerals, 143
Minerals Management Service, 38, 55
Minnesota, 7, 40
minority, viii, 147
missions, viii, 8, 9, 46, 48, 75, 102, 110, 125, 143
Missouri, 24
mixing, 23
MMS, 39, 55, 56, 57
modeling, 20, 21, 129, 142
models, 21, 90, 133
molecules, 46
morning, 137
motors, 71, 77
movement, 31, 98, 99
MTBE, 67
multiple factors, 143
municipal solid waste, 41, 42, 46, 117, 134
municipal solid waste), 42
mutations, 62

N

NAS, 87, 133
nation, 31, 32, 85, 89, 90, 105, 135
National Academy of Sciences, 21, 85, 87, 133
National Academy of Sciences (NAS), 87, 133
National Institute of Standards and Technology, 94
national product, 104
National Research Council, 65
National Science Foundation, 24
national security, 149
National Security Council, 90
National Wildlife Refuge, 66, 67
natural, vii, 2, 6, 7, 10, 11, 12, 13, 23, 26, 29, 42, 45, 49, 72, 73, 79, 80, 82, 88, 91, 99, 108, 110, 130
natural disasters, 80
natural gas, vii, 23, 29, 42, 45, 49, 72, 73, 82, 91, 99, 108, 110
Near East, 50
Nebraska, 16, 112
negative outcomes, 14
negotiating, 81
net metering, 43, 44
Netherlands, 16
network, 47
neurological disorder, 62
Nevada, 26
New Jersey, 44
New Mexico, 25, 36
New York, 3, 16, 35, 44, 61, 112, 154
New York Times, 154
newsletters, 54
next generation, 131, 132, 139, 145
nitrogen, 62, 108, 152
nitrous oxide, 47, 87
non-native, 107
normal, 96, 110, 154
North America, 12, 44
nuclear, 31, 32, 33, 41, 90
nuclear power, 31
nuclear technology, 91
nutrition, 111, 149

O

obligation, 90
obligations, 81, 128, 130
OCs, 46
OECD, 123

Office of Management and Budget, 82, 96
Office of Science and Technology Policy, 66
offshore, 30, 38, 55, 56, 57
oil, vii, 1, 3, 4, 7, 19, 22, 23, 33, 36, 46, 47, 48, 55, 62, 70, 71, 72, 73, 84, 88, 89, 91, 95, 96, 98, 100, 101, 102, 103, 107, 112, 113, 118, 121, 127, 130, 131, 135, 143, 148, 150, 155
oils, 62, 99, 101
oilseed, 154
OMB, 82
omnibus, vii, 69, 70, 72, 73
online, 3, 54, 55, 63, 121
operator, 40
organic, 17, 19, 46
organic compounds, 46
organization, 3
organizations, viii, 19, 38, 39, 129, 147
output method, 115
oxide, 62
oxides, 87
ozone, 62, 129, 152

P

Pacific, 36, 37, 50
palm oil, 4, 7, 143
paper, 60, 66, 121, 122, 123
Paper, 25, 45, 119, 123
parallel processing, 139
Paris, 123
particulate matter, 129
partnership, 11, 32, 41, 47, 50, 90
partnerships, 87
passenger, 74, 86, 108
passive, 52
pasture, 8, 152
pathways, 21
peak demand, 45
peat, 8, 143
pedestrians, 92
penalties, 74, 89
penalty, 143
Pennsylvania, 44
per capita, 144
per capita income, 144
perception, 45
performance, 1, 2, 4, 5, 6, 17, 18, 19, 22, 23, 24, 40, 47, 78, 83, 84, 94, 128, 129, 141, 143
pesticides, 106
petroleum, vii, viii, 7, 15, 47, 61, 74, 82, 88, 89, 97, 98, 99, 101, 102, 107, 108, 109, 116, 118, 121, 126, 128, 130, 131, 132, 138, 139, 140
Petroleum, 61, 62, 123
petroleum products, 89, 116
phone, 57
phosphorous, 152
photosynthesis, 107
photovoltaic, 36, 81, 82
photovoltaics, 31
pipelines, 76, 116, 123, 132, 138
planning, 3, 22, 41, 79, 83
plants, 16, 37, 42, 45, 47, 75, 83, 105, 107, 110, 115, 119, 134, 144
play, vii, 13, 29, 48, 99, 101, 106, 116, 134
plug-in, 37, 74
policymakers, 102, 120, 152
politics, 3, 22
pollutant, 20, 101
pollutants, 49
pollution, vii, 1, 3, 5, 6, 8, 9, 15, 18, 23, 29, 32, 49, 63, 152
pools, 94
poor, 147, 149, 153
population, 143, 144
population growth, 143, 144
pork, 113, 114
portfolio, 32, 35, 44, 58, 59, 72, 73, 138
poultry, 34, 41, 113, 130, 150
poverty, 149, 154
power, vii, 29, 31, 33, 35, 36, 38, 43, 44, 45, 47, 49, 52, 53, 57, 59, 67, 71, 77, 83, 86, 91, 94, 96, 106, 140, 142
power plant, 38, 45, 86
power plants, 38, 86
pragmatic, 15
premium, 44
President Bush, 20, 35, 48, 61, 99, 132

pressure, 12, 14, 104
preventive, 14
price effect, 113
Price-Anderson Act, 90
prices, vii, viii, 1, 2, 4, 8, 10, 17, 29, 30, 32, 33, 44, 45, 48, 60, 97, 98, 101, 104, 108, 109, 111, 112, 113, 114, 115, 116, 119, 130, 138, 144, 147, 148, 149, 150, 151, 152, 153, 154
private, 15, 16, 18, 19, 33, 55, 81, 87, 94, 135, 139
private sector, 15, 16, 18, 33, 55, 135, 139
private sector investment, 16, 139
producers, 4, 17, 35, 37, 43, 59, 63, 113, 120, 126, 127, 129, 130, 141, 142, 143, 150, 155
production, vii, viii, 1, 2, 3, 5, 7, 8, 9, 13, 14, 16, 17, 31, 32, 34, 36, 40, 41, 42, 43, 45, 46, 47, 48, 49, 52, 59, 61, 71, 72, 75, 76, 84, 86, 87, 95, 97, 98, 99, 100, 101, 102, 104, 105, 106, 107, 108, 109, 110, 112, 113, 114, 115, 116, 117, 118, 122, 127, 128, 129, 130, 132, 133, 134, 137, 138, 139, 140, 141, 142, 143, 144, 147, 148, 149, 150, 151, 152, 153, 154
production costs, 105, 109
productivity, 138, 139, 140, 141, 153
profit, 113
profit margin, 113
profits, 111
program, 6, 30, 32, 33, 37, 42, 50, 52, 53, 60, 63, 70, 73, 74, 75, 76, 77, 78, 79, 80, 81, 83, 84, 85, 86, 87, 88, 91, 92, 93, 94, 114, 125, 126, 127, 129, 130, 136, 142, 149
promote, vii, viii, 5, 36, 50, 78, 83, 84, 89, 90, 97, 98, 102, 108, 109, 118
proposition, 13
protection, 11, 93, 94, 140
protein, 18, 112, 113
protocol, 4, 20
protocols, 94
PSD, 122
public, 1, 3, 5, 13, 15, 17, 24, 44, 53, 55, 56, 63, 64, 78, 80, 84, 87, 88, 90, 94, 102, 130
public education, 94
public health, 1, 3, 5, 56, 87
public interest, 55, 102
public policy, 102
public resources, 13, 15
pumping, 81
pumps, 79, 117

R

R&D, 30, 33, 35, 36, 48, 51, 52, 55, 64, 65, 67, 75, 76, 84, 85, 86, 87, 133, 134, 142
radar, 30, 36, 39, 40, 57
Radiation, 125
radio, 39
rail, 116
rain, 32, 49
rainforest, 7, 9
RandD, 18
range, 105, 110, 112, 126, 129, 130, 135, 141
raw material, 116
reading, 20
real estate, 12
real time, 59
reality, 17, 105
recovery, 79, 81, 84
recreational, 39
recruiting, 36
recycling, 85
reduction, 2, 5, 6, 14, 15, 18, 23, 40, 49, 63, 74, 78, 81, 103, 110, 121, 133, 142, 150
refineries, 9
refiners, 116, 126, 129, 132
refinery capacity, 89
reforms, 153
refrigeration, 86
regional, 59, 77, 83, 111, 113
regulation, 17, 18, 21, 128
regulations, 2, 3, 17, 18, 19, 20, 21, 22, 23, 75, 104, 126, 127, 129, 130
regulatory framework, 88
rehabilitation, 92
relationship, 9
relationships, 127
reliability, 93, 96
Reliability, 52

renewable energy, vii, 3, 29, 30, 31, 32, 33, 34, 35, 36, 37, 38, 40, 41, 42, 43, 44, 45, 48, 49, 50, 52, 53, 54, 55, 56, 59, 63, 69, 70, 71, 72, 73, 76, 78, 87, 88, 89, 90, 93
Renewable Energy and Energy Conservation Tax Act, 72
Renewable Fuel Standard, i, iii, 2, 42, 64, 70, 71, 75, 97, 101, 103, 104, 105, 131, 132, 147
Renewable Fuel Standard (RFS), 42, 70, 71, 75, 97, 101, 131
Renewable Fuels Standard, 3, 69, 137, 141
renewable resource, 14, 41
research, 2, 15, 32, 41, 47, 64, 74, 75, 76, 79, 80, 85, 90, 105, 107, 131, 132, 133, 134
research and development, 74, 131, 134
Research and Development, 61, 75, 83, 122
Research and Development (R&D), 75
researchers, 135
reservation, 92
reservoir, 84
reservoirs, 13, 84
residential, 33, 34, 52, 69, 71, 77, 94
residues, 10, 17, 18, 100, 120, 129, 134
resilience, 13
resistance, 26
resolution, 54
resource availability, 41, 47
resources, 13, 15, 31, 32, 36, 41, 43, 46, 71, 81, 84, 87, 88, 89, 93, 96, 132, 133, 134, 144, 152
respiratory, 62
responsibilities, 90
restaurant, 47
restructuring, 34, 44, 59, 73, 128
retail, 41, 44, 47, 59, 71, 96, 111, 113, 117, 149, 154
retirees, 149
revenue, 35, 42, 72
rewards, 23
RFS, viii, 2, 3, 4, 5, 6, 7, 9, 10, 11, 12, 15, 16, 17, 18, 19, 20, 21, 22, 23, 24, 42, 48, 62, 69, 71, 73, 75, 98, 101, 102, 103, 104, 105, 108, 109, 110, 111, 113, 118, 119, 120, 123, 125, 126, 127, 130, 131, 132, 133, 135, 138, 141, 143, 144, 152
rice, 144
rights-of-way, 56, 88
risk, 1, 3, 11, 13, 14, 15, 24, 102, 114, 141, 142, 144, 149, 150
risks, 13, 15, 20, 40
rivers, 152
runaway, viii, 147
runoff, 18, 79, 107
rural, vii, viii, 23, 37, 76, 97, 98, 110, 115, 116
rural areas, 116
rural communities, 76

S

Safe Drinking Water Act, 87
safeguards, vii, 1, 2, 3, 5, 9, 10, 11, 17, 18, 22, 24
safety, 30, 39, 86, 87, 94
sales, 59, 71, 115
savings, 36, 79, 81, 86, 94, 138
SBA, 93
scalable, 14
scaling, 132
school, 53, 79, 114, 149
scientists, 3, 105, 135
second generation, 15, 16, 139
Secretary of Defense, 50
Secretary of State, 90, 91
security, viii, 1, 63, 80, 90, 91, 94, 101, 102, 108, 109, 114, 118, 120, 125, 129, 131, 135, 138, 147, 149
Security Council, 90
seed, viii, 137
seeds, viii, 118, 137
Self, 60
seller, viii, 137
senate, 57, 60, 63, 64
Senate, 1, 23, 29, 30, 36, 37, 41, 42, 43, 44, 45, 49, 50, 51, 57, 63, 64, 71, 72, 73, 96, 98, 99, 102, 108, 109, 110, 116, 117, 118, 119, 123, 125, 131, 137, 147
SEPA, 68

September 11, 123
series, 89
services, 3, 13, 35, 59, 83, 88, 91
sewage, 42
shape, 23
sharing, 83
shipping, 92, 116
shock, 33
short run, 114
short-term, 59
sign, 23
signals, 20, 39, 141
silicon, 36
sine, 67
sites, 14, 39, 84, 88
skeptics, 105
Small Business Act, 93
Small Business Administration, 93
small firms, 93
smog, 49
SO2, 86
Social Security, 149
soft drinks, 113
soil, 1, 5, 7, 9, 13, 14, 18, 107, 111, 139
soil erosion, 18, 107, 111
solar, 29, 31, 32, 34, 35, 36, 41, 42, 43, 44, 45, 52, 54, 55, 63, 82, 83
solar energy, 29, 32, 41, 52, 63
solid state, 86
solid waste, 41, 46, 117, 134
solvency, 95
Sorghum, 100
South Africa, 140
South Asia, 50
South Dakota, 40
Southeast Asia, 143
soy, 8, 18
soy bean, 8, 18
soybean, viii, 62, 101, 111, 112, 113, 137, 140, 148, 150, 154
soybean seed, viii, 137
soybeans, 8, 18, 98, 99, 111, 113, 114, 130, 134, 148, 154
species, 11, 12, 13, 39, 107, 113
speculation, 143, 153
speed, 105
spin, 38
spinoff, 115
stability, 60
stages, 7, 128
stakeholder, 129
stakeholders, 126, 129, 130
Standard and Poor's, 65
standards, viii, 2, 5, 6, 7, 9, 10, 17, 18, 19, 22, 23, 55, 56, 58, 69, 70, 71, 72, 73, 77, 78, 80, 94, 125, 127, 128, 135, 141, 143, 144
starch, viii, 75, 98, 101, 103, 105, 106, 118, 123, 137
starches, 46
State of the Union, 65, 99, 132
statistics, 58, 119, 154
statutory, 130
steel, 138, 144
storage, 14, 15, 63, 81, 83, 85, 86, 87, 88, 143
strain, 116
strains, 138
strategic planning, 83
strategies, 4, 79, 91, 94, 118, 141
stressors, 80
students, 80, 150
subsidies, 70, 71, 72, 73, 102, 105, 111, 113, 115, 151, 152, 153
subsidization, 9
subsidy, 34
subsistence, 143
subsistence farming, 143
substitutes, 101, 109, 115, 116
substitution, 48
sugar, 46, 99, 100, 101, 106, 110, 118
sugar cane, 101, 110
sugarcane, 98, 99, 108, 109, 118
sugars, 46
sulfur, 62, 86
sulfur dioxide, 62, 86
summer, 45, 138
suppliers, 41, 59, 71, 91
supply, vii, viii, 9, 14, 17, 18, 31, 37, 48, 97, 98, 99, 103, 104, 106, 108, 116, 120, 132, 134, 138, 144, 151, 154
supply chain, 132

surging, 150
survivors, 149
sustainability, 80, 93, 133, 141, 142, 144, 153
switching, 37
Switzerland, 119, 123
systems, 11, 15, 30, 32, 37, 39, 40, 44, 78, 79, 83, 84, 85, 87, 91, 94, 116, 127, 130, 132

T

Taiwan, 115
tanks, 117
tar, 63
targets, 40, 86, 135, 141
tariff, 1, 4, 23, 59, 100, 119, 153
tariffs, 4, 151, 152
tax credit, 1, 4, 5, 23, 32, 33, 34, 35, 40, 43, 45, 48, 52, 53, 63, 72, 104, 153
tax credits, 4, 23, 33, 34, 153
tax incentive, 52, 72, 73
tax incentives, 52, 72, 73
taxes, 41, 95
technical assistance, 79
technology, vii, 1, 4, 15, 17, 19, 22, 24, 29, 34, 43, 46, 47, 48, 52, 61, 65, 84, 85, 86, 91, 93, 94, 102, 105, 117, 126, 134, 135, 138, 141, 142, 153
technology transfer, 84
Tennessee, 16
terminals, 116
testimony, 9, 24, 126, 137
Texas, 35, 38, 44, 64, 130, 133
thermal energy, 34, 55, 81, 83
threat, 11, 12, 114
threatened, 11, 73
threats, 12, 108
threshold, 128, 141
thresholds, 129
tides, 85
timber, 26, 143
time, 4, 6, 10, 14, 18, 32, 33, 43, 59, 79, 96, 108, 120, 127, 128, 129, 139, 142, 152, 153
timing, 22
title, 77, 82, 88, 89, 91, 94
Title III, 49, 50, 77
total energy, 34, 78, 107
tourism, 38, 39
tourist, 39
toxic, 102
tracking, 19, 127
trade, 9, 23, 102, 107, 114, 127, 142
trade policies, 9
trade-off, 107
trading, 3, 104
traffic, 13, 40
training, 91, 94
traits, 86
trans, 111
transactions, 59
transfer, 84
transition, 4, 140
transitions, 141
transmission, vii, 29, 35, 44, 59, 83, 85, 93
transparency, 91
transport, 8, 82, 107, 116
transportation, vii, viii, 1, 3, 8, 14, 22, 23, 46, 47, 48, 75, 76, 91, 92, 97, 98, 100, 102, 106, 108, 116, 125, 126, 132, 134, 135, 138, 139, 142, 152
Treasury, 74, 95
Treasury Department, 95
trees, 7, 10, 12, 13, 46, 55, 76, 99, 106, 134
trend, 54
tribes, 88
tropical forest, 143
tropical forests, 143
trucks, 69, 70, 74, 116, 117
trust, 95
trust fund, 95

U

U.S. Agency for International Development, 50, 90
U.S. Department of Agriculture, 37
U.S. Department of Agriculture (USDA), 37
U.S. economy, 115
U.S. Geological Survey, 89
U.S. Treasury, 74
ubiquitous, 117

UN, 148, 149, 153
UNCED, 32
uncertainty, 14, 20, 60, 105
unemployment, 95
UNEP, 63
United Kingdom, 30, 38, 40
United Nations, 32, 54, 63, 66, 67
United Nations Environment Program, 63
United States, 1, 9, 11, 12, 25, 27, 38, 41, 45, 47, 63, 66, 79, 81, 84, 88, 89, 90, 92, 97, 98, 99, 100, 101, 104, 105, 108, 112, 114, 116, 117, 118, 119, 120, 121, 122, 123, 125, 131, 132, 135
universities, 76, 80, 133
unobligated balances, 82, 92
upload, 123
urban areas, 37
USAID, 50, 90
USDA, 19, 25, 37, 47, 48, 49, 53, 65, 104, 106, 112, 119, 120, 121, 122, 129, 130, 132, 133, 141, 148, 154, 155, 156

V

validation, 133
values, 7, 13, 19, 20, 21, 151
vegetable oil, 47, 100, 148, 150
vehicles, vii, 1, 3, 42, 53, 74, 76, 82, 85, 86, 91, 101, 117, 118, 127, 133
venture capital, 16, 93, 102
Vice President, 137
volatility, 138
vulnerability, 44

W

wage rate, 83
wages, 95
Wales, 41
Wall Street Journal, 123, 154
Washington Post, 57
wastes, 55, 134
water, vii, 1, 3, 5, 12, 17, 18, 29, 31, 36, 38, 39, 41, 46, 49, 78, 81, 82, 129, 132, 133, 134, 140, 144, 152
water heater, 36
water quality, 129, 134
water resources, 81
waterways, 152
Western Hemisphere, 50
wheat, 111, 114, 134, 144, 150
White House, 61, 65, 73
wholesale, 19, 35, 44, 59, 109, 112
wholesalers, 89
wilderness, 26
wildfire, 11, 14, 24
wildland, 26
wildlife, 1, 3, 5, 10, 11, 12, 13, 18, 56, 152
wind, 30, 31, 32, 34, 35, 36, 38, 39, 40, 41, 42, 43, 44, 45, 50, 53, 56, 57, 65, 87
wind farm, 30, 35, 38, 40, 45, 57
wind turbines, 30, 31, 38, 39, 40
winter, 56
wintertime, 47
Wisconsin, 40, 135
wood, 12, 18, 31, 45, 55, 134, 135, 139
wood products, 12
wood waste, 18, 45, 55, 134, 135
woods, 14
workers, 149
World Bank, 48, 136, 148, 149, 153

Y

yield, 18, 32, 121, 140